Sydney

Christina DeMarco, Brian Riera

and

Peter Spearritt

AIRD BOOKS
Melbourne

Aird Books Pty Ltd
PO Box 122
Flemington, Vic. 3031
Phone (03) 376 4461

First published by Aird Books in 1991
Copyright Part 2, History © Peter Spearritt 1991
Copyright all other text © Christina DeMarco & Brian Riera
Copyright maps and illustrations © Aird Books 1991

National Library of Australia
Cataloguing-in-publication data

DeMarco, Christina.
  Aird's guide to Sydney.

  Bibliography.
  Includes index.
  ISBN 0 947214 10 0.

  1. Sydney (N.S.W.) – Description – 1976-   –Guidebooks. I. Riera, Brian. II. Spearritt, Peter,
  1949-   . III. Title. IV. Title: Guide to Sydney.

919.4410463

Cover design by Lin Tobias
Lay-out by David Spratt
Maps by Country Cartographics
Illustrations by Greg Jorgenson
Printed in Australia by Australian Print Group,
  Maryborough, Victoria

Acknowledgments

Christina and Brian would like to thank Ronald and Peggy Nott,
who have helped them discover many of Sydney's special places
over the years.

The authors and publishers would like to thank the follow-
ing people and organisations for providing the photographs in
this book: Ross Brownscombe, the National Library of Australia
(Canberra), Tourism Australia (Sydney), Darling Harbour
Authority (Sydney), the National Maritime Museum, Northside
Productions (Melbourne), and Pittwater Yacht Charter (Lovett
Bay, NSW).

Acknowledgment is given with the photographs, as ap-
propriate; photographs without a credit have been supplied by
the authors.

Readers' comments

Any criticisms, suggestions, and constructive comments on
this Guide by travellers and other interested parties will be
gratefully received by the author and publishers. Please ad-
dress all correspondence to:
  The Editor
  Aird's Guide to Sydney
  PO Box 122, Flemington, Victoria 3031
  Australia

Contents

Authors' note

Aird's Guide to Sydney combines the talents of town planners and travellers Christina DeMarco and Brian Riera. Based in Sydney, they have worked, lived, and travelled in several of the world's great cities but have a special affection for Sydney.

Peter Spearritt is currently Director of the National Centre for Australian Studies at Monash University in Melbourne. He has written extensively on the social history of Sydney, where he lived for many years, and his books include *Sydney since the twenties*, *The Sydney Harbour Bridge: a life*, and—co-authored with Christina DeMarco—*Planning Sydney's future*.

PART I
Information

Facts for overseas visitors

Travel papers

Every visitor to Australia requires a valid passport. All nationalities, with the sole exception of New Zealand citizens, also require a visa. For most people this will mean a tourist visa.

Tourist visas must be obtained before you embark for Australia. They are available from your nearest Australian Consulate.

Working holiday visas are also available to people between the ages of 17 and 25 for a period of not more than one year. They are available only to citizens of the UK, Canada, The Netherlands and Japan.

Visitors travelling from countries with endemic diseases such as yellow fever, cholera, or typhoid, are required to have inoculation certificates.

Quarantine

Australia is particularly vulnerable to outside pests and diseases and consequently quarantine laws are strictly enforced. Do not import any animal or vegetable matter, living or dead. If you have any doubts be sure to declare any items to customs on entry.

Your health in Australia

Australia is a particularly clean and healthy country. There is no need to worry about the quality of the drinking water or contagious diseases.

However don't overlook the sun as a potential health hazard. Be careful of the sun, especially if you are coming from the northern winter. The fierceness of the sun can surprise and deceive on Sydney's beaches where there are often pleasant breezes. Sydney is probably the best place to buy your sun screen and zinc cream so no need to bring it with you. And don't save it just for the beach. You may have as much chance of sunburn walking the hot streets of the city.

Money

The Australian Dollar is the unit of currency. It is free floating, and can fluctuate even within the short time of your holiday. At the time of publication, one $A was worth about .78 US ($); .40 UK (stg); 1.15 Deutschmarks; 1.30 NZ ($); and 102 Japanese yen.

Travellers cheques are widely accepted in Australian banks and major hotels. Many banks charge a commission but this varies as do the exchange rates. There's not a lot of difference between banks, but expect currency exchange offices to be quite a bit more expensive. The Westpac Bank operates a *Bureau de Change* at the international air terminal.

Mastercard, Amex, and Diners. The Australian banks usually provide services for Visa or Mastercard. Cards are a good way of collecting your Australian cash if you prefer not to carry it around in cheques. You may pay a little more in commission but the convenience is probably well worth it. Exchange rates are generally better for travellers cheques than for cash.

Organising your trip

For the beach and hot sunny weather December to March are the best times to visit Sydney. November and April are usually beautiful in Sydney but the water is not quite so pleasant. If you're travelling to other parts of Australia, the wet season is November to March in northern Queensland. This is also the hot season in Central Australia and Ayers Rock.

Winter is cooler though still mild and is perfect walking and touring weather. The Australian winter is the best time to visit Northern Queensland and Ayers Rock in Central Australia.

For most general tourist trips booking ahead is usually unnecessary unless a particular hotel is favoured. Many hotels offer a better price if booked overseas through an agent or as part of a package, so check out these opportunities well ahead if you are sure of your itinerary.

International ticket holders are entitled to a 25% discount off the price of tickets on most domestic airlines. This can help make a dent in the expensive domestic ticket prices.

Arriving in Sydney

Sydney's Kingsford-Smith Airport

The airport is conveniently located about 10 kilometres south-west of the harbour and the centre of town. To *change money* there is a bank on the arrivals level as you leave the customs hall. Rates may be a little higher here than in the city but not seriously so. The *Tourist Information Service* is also available on the arrivals level. They should be able to find you accommodation to meet your budget and there are brochures by the desk advertising hotels and backpacker accommodation. They should also be able to book buses, planes, etc. for you, and they have free maps of the city. There are some hotel freephones in this area. *Lockers* are available on the arrivals and departures levels of the terminal at a cost of $1 for the first day and $2 for every additional day. Hertz, Avis and Budget *rental cars* can be found on the arrivals level and are open at 6 a.m.

Getting into the City

Because the airport is so close to the centre of the city, this is a pretty easy task. Here are the choices:

Taxis to the city centre cost about $15 and usually take about 15 minutes, except during peak traffic hours.

Public buses are marked "City Airport Express" and will take you to the city for $4. They stop at the main city railway stations including Central Station and Circular Quay and the domestic terminals on the way. The trip to the city centre takes between 15 and 30 minutes. They run about every 20 minutes, 7 days a week but don't always keep to schedule. To return to the airport, there are bus stops marked through the city on George Street and at Central Station and Circular Quay.

Private mini buses provide the same service as the public bus at the same cost of $4. They also serve the Kings Cross area, North Sydney and the Eastern Suburbs. They usually wait until

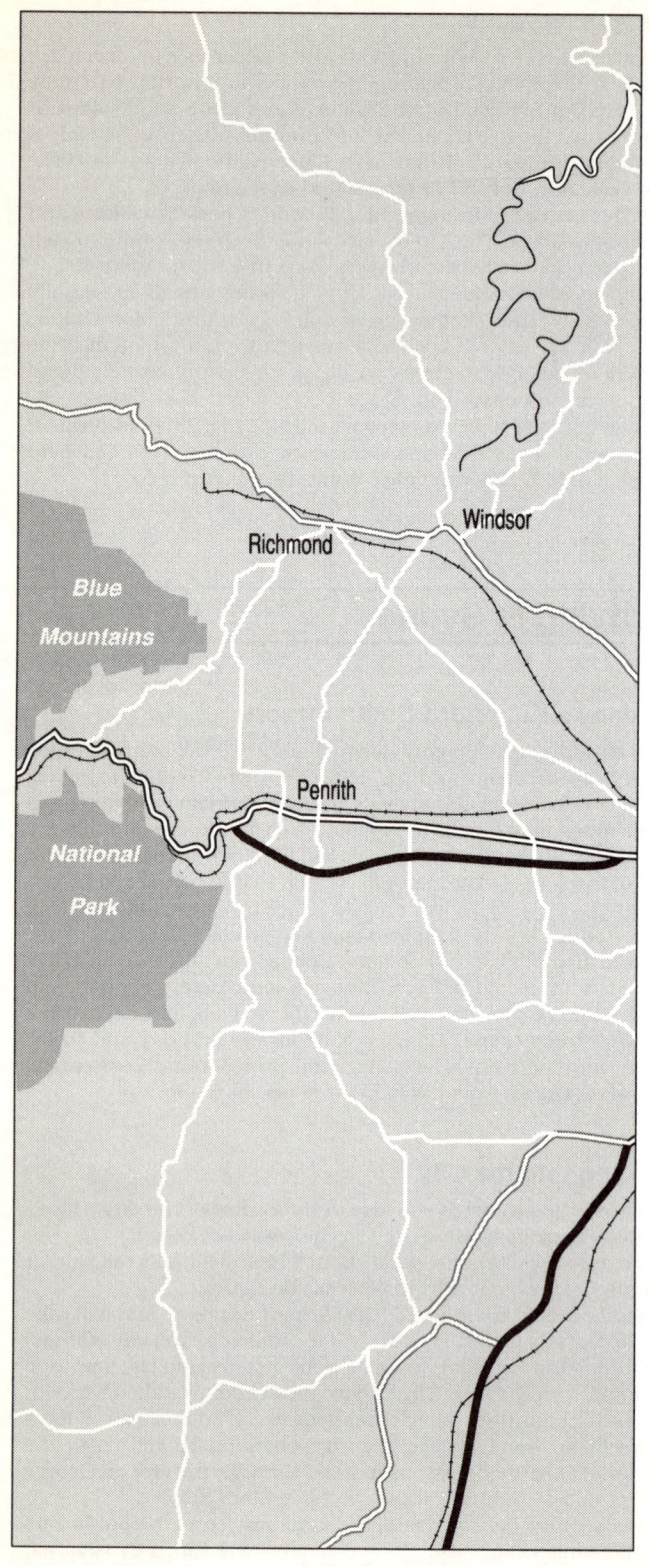

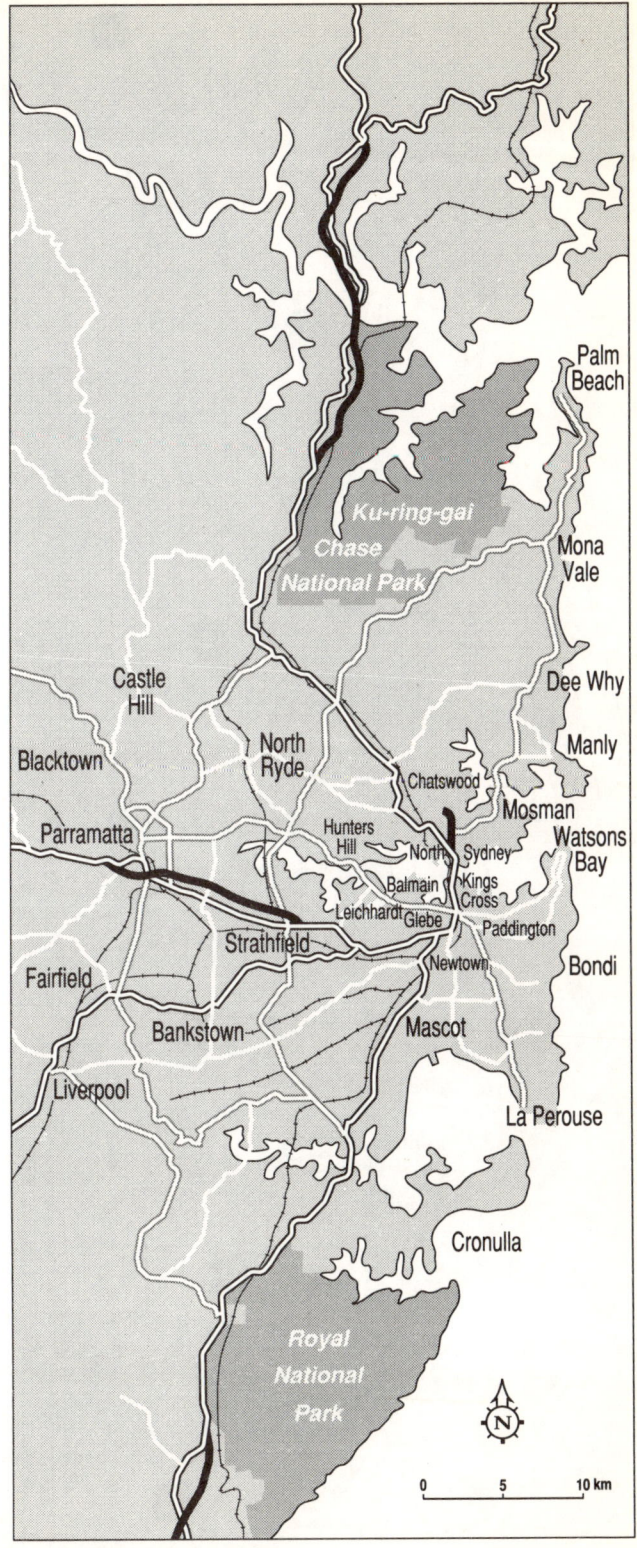

Sydney region

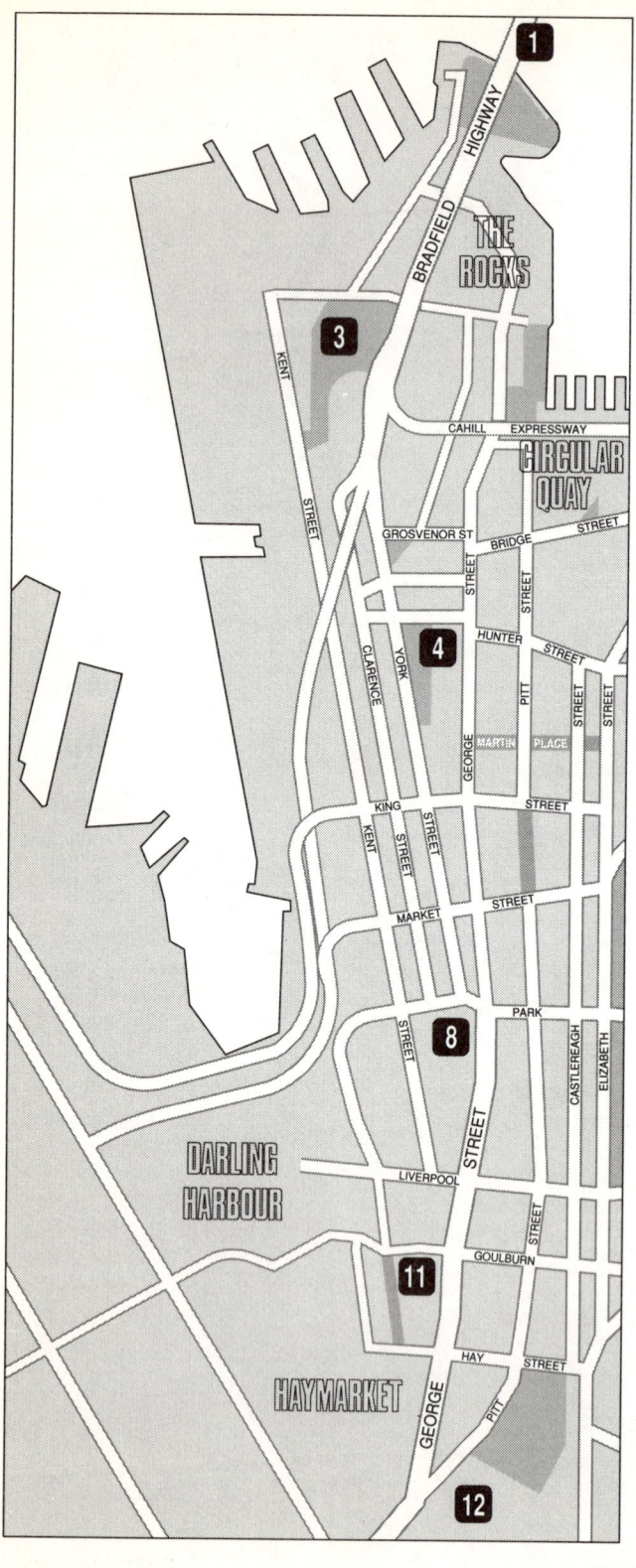

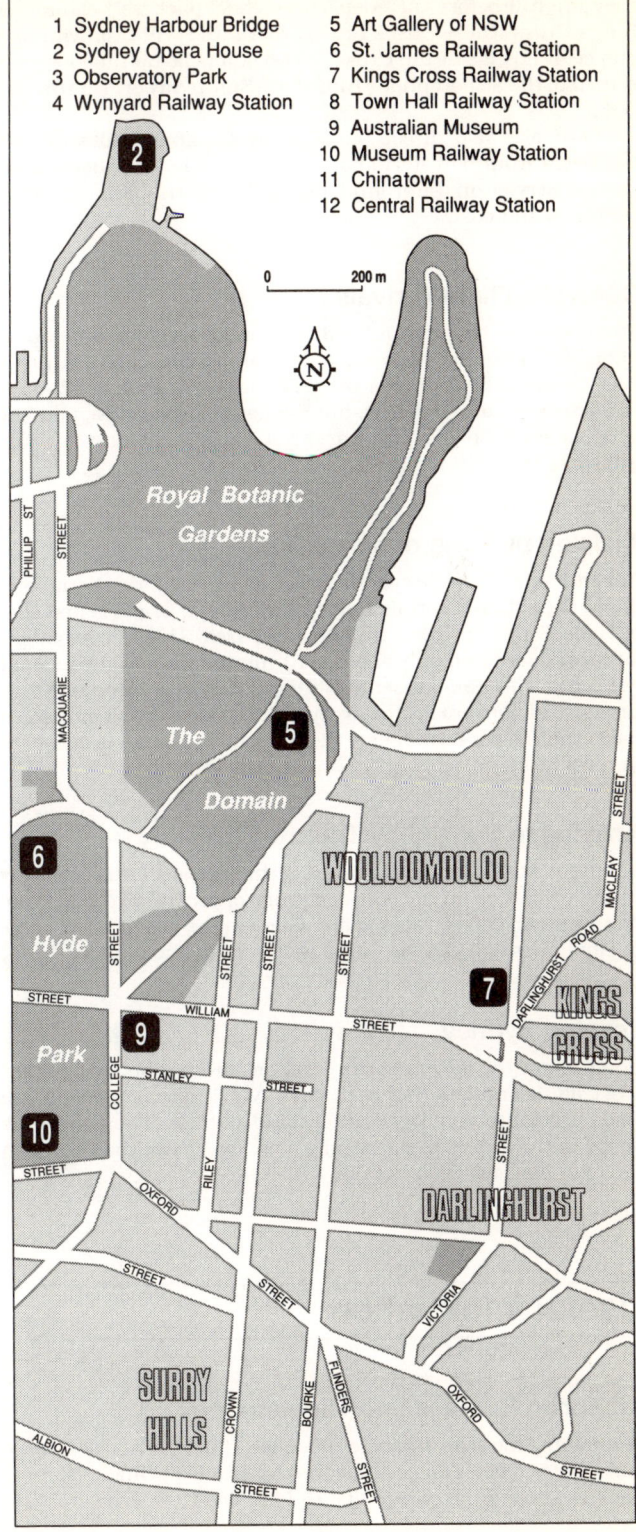

1 Sydney Harbour Bridge
2 Sydney Opera House
3 Observatory Park
4 Wynyard Railway Station
5 Art Gallery of NSW
6 St. James Railway Station
7 Kings Cross Railway Station
8 Town Hall Railway Station
9 Australian Museum
10 Museum Railway Station
11 Chinatown
12 Central Railway Station

0   200 m

N

Royal Botanic Gardens

PHILLIP ST

STREET

MACQUARIE

The

Domain

WOOLLOOMOOLOO

MACLEAY

STREET

Hyde

STREET

STREET

STREET

DARLINGHURST ROAD

KINGS

STREET

CROSS

STREET

Park

WILLIAM

STREET

DARLINGHURST

COLLEGE

STANLEY

STREET

STREET

VICTORIA

OXFORD

RILEY

STREET

SURRY

CROWN

BOURKE

FLINDERS

OXFORD

HILLS

ALBION

STREET

STREET

STREET

Central Sydney

they are full before they depart. For backpackers heading for Kings Cross, these buses are excellent and will drop you at a choice of hostels at the Cross. To book a pick-up time for your return to the airport phone Kingsford-Smith Airport Bus on 667 0663.

There are also direct buses to Wollongong, Gosford and Canberra from both terminals. Check at the Travellers Information Service on the ground floor of the international terminal for times and cost.

Domestic Flight Arrivals

The domestic terminal is at the same airport but is about 5 kilometres from the international terminal. The Airport Express Bus service runs between the terminals at a cost of $2.40 or you can take a taxi. There are domestic airline desks on the arrivals level so it's best to check with them first for any information about flights.

Arriving by Long-distance Bus

Ansett Pioneer and Greyhound are the two major national bus lines. They share a terminal on Oxford Street, which is just east of Hyde Park and the main shopping area. This is convenient for local buses and only about a 10-minute walk to the nearest suburban rail line (Museum Station). It is also close to Kings Cross, and convenient for many hotels. Other bus companies have their own terminals. The Deluxe terminal is close to Central Station.

Arriving in Sydney by Train

All interstate and intercity trains terminate at Central Railway Station which is on the southern edge of central Sydney. From here you can catch trains to anywhere in the city. A good range of city buses can also be found just outside the station.

Leaving Sydney

Don't forget the $20 departure tax on international flights for every adult and child 12 years and over. There is a post office for last minute posting of letters and parcels. The Duty Free shops at the airport are generally more expensive than those in the city. See Part 8 for details.

Useful information

American Express and Thomas Cook

The main American Express office is on the corner of George and King Streets, just south of the General Post Office. Thomas Cook is around the corner at 175 Pitt Street.

Baby Changing Rooms

The major department stores in the city centre, David Jones' and Grace Bros have comfortable facilities, usually called mother's rooms. There is also a changing room in Hyde Park behind St James Station and at the Traveller's Aid at the Central Railway Station.

Climate

Sydney has about 120 cm (48 inches) of rain a year. Often when it rains, it rains heavily. But apart from a few ferocious thunderstorms, Sydney's climate is marvellous. The winter months are mild and the summer is hot but rarely unpleasantly so. Still, you should be aware that temperatures vary noticeably from the east to the west of Sydney. The western parts are cooler in winter and often 4^0 or 5^0 C hotter in summer. The Blue Mountains rarely have snow, but are often cold and wet in winter.

The charts below give a quick summary of average minimum and maximum temperatures as well as rainfall and sea temperature. To convert to Fahrenheit, divide by 5, multiply by 9, and add 32; a short cut is to add 15 and then double.

Despite the pleasant winter temperatures, the sea is too cold for most Sydneysiders between May and November.

One Sydney special is the wind known as the Southerly Buster. On a hot and humid summer day the Buster can arrive very suddenly and drop temperatures by 10^0 C and more in less than 30 minutes.

	Jan	Feb	Mar	Apr	May	Jun	Jul	Aug	Sep	Oct	Nov	Dec
Daily Maximum (^{o}C)	26	26	25	22	19	17	16	18	20	22	24	25
Daily Minimum (^{o}C)	19	19	17	15	11	9	8	9	11	13	16	17
Rainfall (cm)	10	11	14	13	12	13	10	8	7	8	8	8
Sea Temperature (^{o}C)	21	22	22	21	19	18	17	16	17	17	18	20

Source: Bureau of Meteorology, NSW Regional Office

Crime and Safety

Sydney with a population of over 3 million people has the normal attractions and problems of big cities. Crime is not the problem it is in American cities, even so, there are parts of the city where you should be on your guard. Kings Cross, the notorious red-light district is the obvious example. Pickpockets, bag snatchers and muggers frequent this area so use normal caution. This doesn't mean that you should be afraid to walk in this or neighbouring districts at night. In fact you will find it thronged with people at all hours of the day and night.

Other areas, such as Central Railway Station, Hyde Park, Oxford Street and Darlinghurst, while safe during the day, might be less safe late at night. Be cautious, and avoid the quieter spots at night.

Currency Exchange

Banking hours are generally Monday to Thursday 9.30 a.m. to 4 p.m. and Friday 9.30 a.m. to 5 p.m. Most large hotels will change travellers cheques although their exchange rate may not be as good as at the bank.

Currency exchange offices are usually places to avoid if you can find a bank open. Still, if you need money on a Saturday or Sunday, try the currency exchange at *Circular Quay*, under the railway bridge opposite Jetty 6. Opening hours are 8 a.m. to 8 p.m. Monday to Saturday, and 9 a.m. to 6 p.m. on Sundays. In *Kings Cross* the Thomas Cook currency exchange is open Monday to Friday, 8.45 a.m. to 5.30 p.m., and on Saturdays 8.45 a.m. to 1 p.m. Thomas Cook also has a currency exchange in the *Queen Victoria Building*, next to the Town Hall, open 9 a.m. to 5.30 p.m. Mondays to Fridays, and till 9 p.m. on Thursdays, 9 a.m. to 4 p.m. on Saturdays, and 11 a.m. to 4 p.m. on Sundays.

Drinking

Sydney's pubs are called hotels and the drinking age is 18 years of age. The usual opening hours are 10 a.m. to 10 or 11 p.m. but closing times can vary. Brasseries, taverns, clubs, wine bars, etc. have different licensing arrangements and many are open until the wee hours of the morning.

The expression BYO seen at many restaurants means that you can "bring your own" alcoholic beverages to the restaurant.

Electricity

Like Europe and most of Asia, Australia's electricity is 240 volts, and 50 cycles. So North American and Japanese appliances will not work without a transformer, unless they have a voltage adjustment switch.

Emergency Services

Emergency: police, fire, ambulance — phone 000

Hospital: outpatients and emergency at Royal Prince Alfred at Camperdown; Sydney Hospital on Macquarie Street has a limited range of services.

Doctors: see "Medical practitioners" in Yellow Pages

Chemists: emergency 24-hour prescription service (drug stores) — phone 438 3333

Dental emergencies: see Dental Emergency Information services and Dentists in Yellow Pages

Life Line — phone 264 2222

Rape Crisis Centre — phone 819 6565

Entertainment

The daily newspapers give a good coverage of what is on at the Opera House, theatres, cinemas, and other venues. The free magazine called *This Week in Sydney*, available in most hotels, at the airport, and at tourist information booths, is a useful introduction.

There is also a 24-hour entertainment hotline, phone 11 688. "Halftix" at Martin Place sells tickets at half price on the day of the performance. See the Entertainment section in Part 8 for more information.

Leaving Luggage

There are coin lockers at Central Railway Station for day storage until 11 p.m. There are also coin lockers at Town Hall and Wynyard railway stations in the city.

Metric Measurements

Australia uses the metric system of measurement. A few useful conversions are listed here. Temperature conversions are listed under climate.

one metre = 1.09 yards (about 39 inches)

one kilometre = 0.62 miles (about 2/3 mile)

To convert kilometres to miles, multiply the number of kilometres by 0.62, i.e. 50 kilometres x 0.62 = 31 miles

1 litre = 1.06 quarts, or 0.26 gallons

1 kilogram = 2.2 pounds

Newspapers

Sydney has two morning papers, a broadsheet, *The Sydney Morning Herald*, and a tabloid, *The Daily Telegraph*. There are also two national morning papers, The *Australian* and *The Australian Financial Review*. The afternoon tabloid is *The Daily Mirror*.

At the large news stands you will be able to find most interstate papers. Overseas newspapers are available at a price. *USA Today* is widely available in the city centre. The newspaper shop at Kings Cross, outside the train station has an excellent selection of interstate and foreign newspapers and foreign magazines.

Postage

Aerogrammes are a good buy at about the same cost as postcard postage and much cheaper than letter postage. If you're sending packages, and cannot wait for sea mail, check out the rates for Economy Air which is quite a bit cheaper than air mail and takes about three weeks to arrive in North America and Europe. The post offices have a wide choice of special envelopes and boxes.

The General Post Office is in the heart of the city on the corner of Martin Place and George Street. Opening hours are from 8.15 a.m. to 5.30 p.m., Monday to Friday and from 8.30 a.m. to 12 noon on Saturday. The poste restante service is here, and generally has a line-up of travellers. The address is, Poste Restante, GPO Martin Place, Sydney 2001. Poste restante services are open till 5 p.m. on weekdays.

Public and School Holidays

Banks, offices and most stores are closed on public holidays. Check opening times for tourist attractions.

- December 25: Christmas Day
- December 26: Boxing Day
- January 1: New Year's Day
- January 26: Australia Day

- March or April: Good Friday, Easter Saturday, Easter Monday
- April 25: ANZAC Day
- Second Monday in June: Queen's Birthday
- August Bank holiday: First Monday only
- Labour Day: First Monday in October

During school vacations the city's attractions will be busier and you may find accommodation a little more difficult to come by. The impact of school holidays, especially in the summer and in the winter break will be most obvious on the coast and at the ski resorts. School holidays are:

- Summer holidays: Christmas to early February
- Easter Break: usually about 10 days
- Winter Break: two weeks end of June/July
- Spring Break: two weeks end September to early October

Shopping Hours

Shopping hours were recently deregulated in Sydney, but still are not as free as in North America. Normal trading hours are 9 a.m. to 5.30 p.m. Monday to Friday, except Thursday when shops are open until 9 p.m. Many shops also stay open until about 7 p.m. on Friday night. On Saturday, shops are open from 8.30 a.m. to 4 p.m. but outside of main shopping areas, many close at noon. On Sundays most shops are closed unless they are in tourist areas such as Circular Quay, Kings Cross, Queen Victoria Building, and Darling Harbour.

Telephones

Look out for red phones in stores. These are for local calls and cost 30 cents. Orange and blue phones are in similar places and are good for long distance calls. The regular green phones in phone booths will handle international calls as well, though many suffer the mysterious public phone malaise which means you may have to search for one in working order. Pay-phones that accept credit cards are becoming more frequent.

Television and Radio

There are five television stations. The ABC Channel 2 is a public TV with no advertising. Also public is SBS, a multicultural station.

There's a wide selection of radio stations, including:

- 2BL ABC 702
- 2FC ABC-National 576
- ABC FM 92.9FM
- 2JJJ ABC-Rock 105.7FM
- 2EA Multicultural 1386
- 2GB 873
- 2MMM 104.9FM

Time Zones

Sydney is in the Eastern time zone which is 10 hours ahead of GMT. Daylight saving takes place between November and March.

Tipping

Australians pride themselves on being an egalitarian lot. Tipping is not necessary, although many people do tip in restaurants to reward good service. For taxis you may make up the bill to the dollar but there is no obligation.

Tourist Information Offices

The Travel Centre of New South Wales is located in the MLC Building, 19 Castlereagh Street near Martin Place, phone 231 4444. Most of the information available on attractions and accommodation is here. The tourist office will also make bookings for travel, accommodation and tours. Opening hours are 9 a.m. to 5 p.m. on weekdays only. There is also a tourist information booth at the Quay, open every day from 9 a.m. to 5 p.m. Outside business hours, try the tourist information service at the airport which is open every day between 6 a.m. and 11 p.m., phone 669 5111.

Videos

Television and video are the same as the European, PAL colour system using 640 lines. So North American video players will not play on Australian TVs. All types of blank video tapes are available in Sydney shops.

Getting around

Sydney's Rail System

Sydney's extensive rail system focuses on the city centre and the Central Railway Station. A section between Central Station and the Harbour Bridge is underground, giving direct access to the major tourist sights and the central shopping and business sections. Two additional underground routes include a loop from Circular Quay back to Central Station, and a route to the eastern suburbs, terminating at Bondi Junction.

Watch out for some surprises. Not all the doors of the trains shut which can be a little unnerving at rush hour, and not all doors open without a tug; the old trains have manually operated doors, which means they are usually open in the summer. The new city and suburban train cars are double-decked, and can be a struggle to get off in the rush hours. Another feature of the system is out of peak services may run only every 15 to 30 minutes so keep that in mind if you've got a schedule to keep. Timetables are available at the railway station. Daily or weekly tickets for unlimited travel are good value. A wide range of

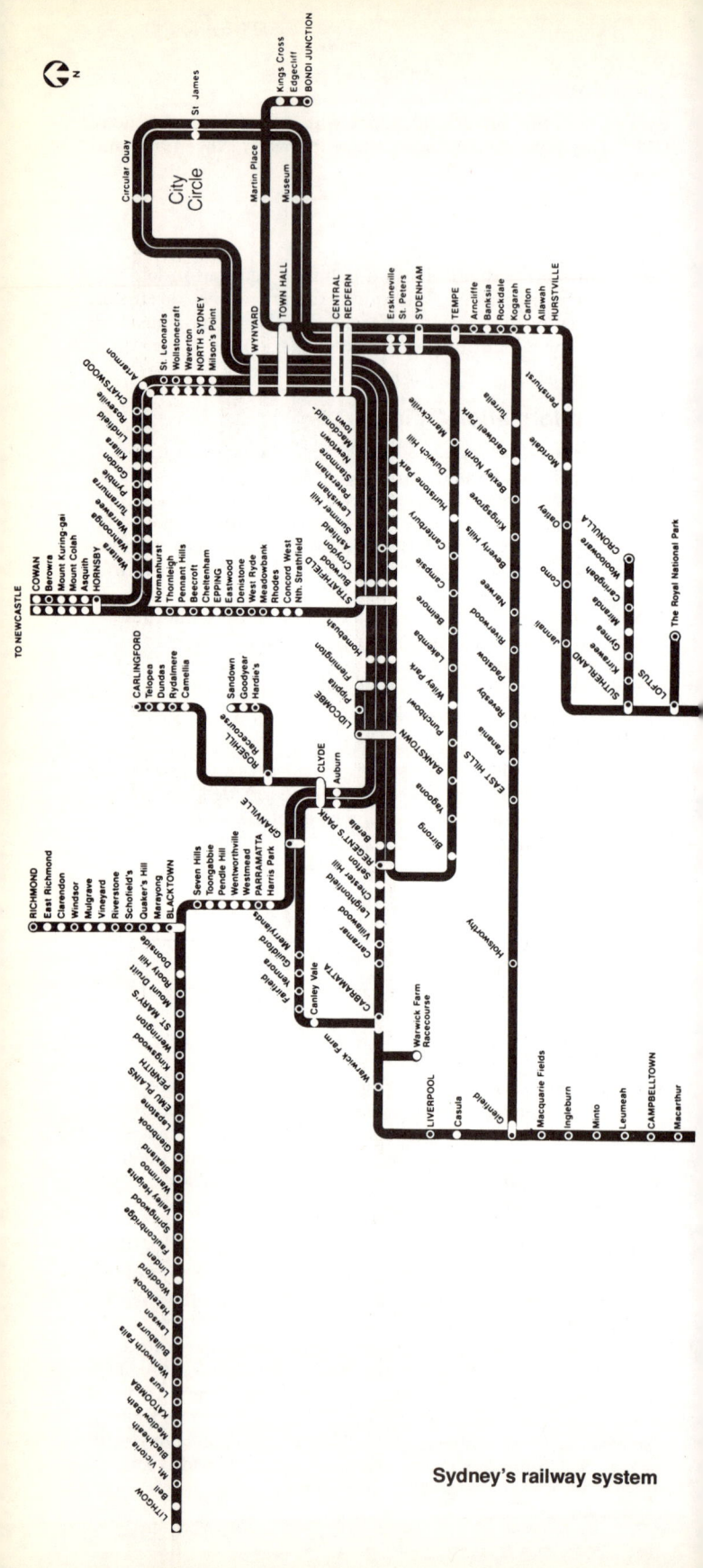

Sydney's railway system

discount and combination tickets are available, described in the section below on combined tickets.

State Rail has an information office downstairs at Queen Victoria Building at Central Station and at Circular Quay. You can book tickets for trips outside Sydney and New South Wales here.

City Buses

Buses can be agonisingly slow in Sydney's congested streets. But for most places in the inner city buses are still a good way to travel. Important pickup points include Circular Quay, Central Station, behind Wynyard Station on York Street, and Bondi Junction.

While drivers will sell single tickets on the bus, you can make significant savings by buying 10-trip tickets or weeklies at newsagents. See Public Transit Discount Tickets, below.

Route maps are useful and available from most newsagents for $2.00. Bus timetables are available from Wynyard Park in Carrington Street and from Circular Quay. Most newsagents stock timetables for buses serving their area.

Some Useful Routes

Route 666 runs every 30 minutes from 10.10 a.m. to 4.40 p.m. Monday to Friday and goes to the Art Gallery with stops en route, including Wynyard Station, Bridge Street (at Pitt Street), and Hyde Park Barracks.

Route 380-2 to Bondi Beach and Centennial Park, from the Quay, or from Elizabeth Street (at Liverpool Street), or take a bus from Bondi Junction.

Route 380 to Oxford Street and Centennial Park, from the Quay.

Route 389 to Paddington (ask for stop), leaves from Circular Quay.

Route 238 to Balmoral Beach, from Taronga Zoo/ Wharf jetty.

Route 190 to Palm Beach, from Avalon, Newport, Narrabeen, from Wynyard Station.

For details on public transport ring 954 4422 between 6 a.m. and 10.00 p.m. daily.

Harbour Ferries

Harbour ferries from Circular Quay cross mainly to the north shore. One of the best and easiest treats in Sydney is a leisurely ferry ride to one of the beautiful bays on the north shore. The Ferry Information Centre will give you any information you need on public ferry routes and timetables, or ring 954 4422.

The major destinations of the public ferries include Neutral Bay, Mosman, Hunters Hill, Balmain, Meadowbank, Taronga Zoo, and Manly. There is a special weekend ferry to Watsons Bay at the South Head of the harbour and it also stops at Taronga Zoo and Rose Bay.

A ferry goes to Darling Harbour every hour on the hour daily. It also calls in at McMahons Point and Balmain, making it a pleasant harbour trip. Cost is $2.40 each way.

The Manly ferry takes 35 minutes and runs every half hour. Hydrofoils, which are being replaced by new Jetcats, depart every half hour to hour and do the trip in 15 minutes. Regular ferries cost $3.20, and the hydrofoil, $4.30.

There is also a ferry every hour up the Parramatta River to Meadowbank. The round trip takes just over an hour and it's an easy and pleasant way to see the river.

A few privately-operated ferries also run from Circular Quay. Hegarty's Ferries cross directly to the other side of the harbour (Kirribilli and McMahons Point).

Water taxis are a quick though expensive way to cross the harbour. You can hail them, but usually it is easier to phone: try 552 2266. They pick up from the normal ferry jetties, as well as from smaller jetties, such as at the Opera House.

Public Transit Discount and Combined Tickets

If you will be using public transport a lot, consider a weekly or daily combined ticket. There is a great variety of tickets, and arrangements change, so you will need to check the details. **Off-peak rail** reduced fare tickets are available for return journeys beginning after 9 a.m., and all day Saturday and Sunday. Bus and ferry tickets can be bought in lots of ten at big discounts, called **MetroTen** and **FerryTen**. Discounted weekly travel passes are also very good value, and are the only tickets that can be used on all buses, trains, and ferries. The **Bus Tripper** offers unlimited travel on buses, but excludes the Airport bus and the Sydney Explorer. The cost is $7 a day.

Sydney Explorer

The tourist bus, the **Sydney Explorer**, will get you around to the major city sights quickly. The fare is $12 for a day's unlimited travel, children and pensioners are $6, and a family ticket for two adults and two or more children is $25.

The big red bus is easy to spot around town, and it runs a 20-kilometre circuit around the city, stopping at 22 tourist spots, including the Opera House, Mrs Macquarie's Chair and the Art Gallery, Kings Cross, Central Railway Station, Chinatown, Darling Harbour, Powerhouse Museum and the Rocks. The Explorer stops are marked by a red sign at the bus stop. An added bonus, you can also use the Explorer ticket on any bus between Central Railway Station, the Quay and the Rocks.

The Explorer runs every 17 minutes between 9.30 a.m. and 5 p.m. and takes about 1 hour 25 minutes to complete the round trip. Buy tickets on the bus, or at the Travel Centre of NSW, at 19 Castlereagh Street, or at the Circular Quay Countrylink Rail Travel Centre. The price of the ticket includes a small guidebook.

Taxis

There are usually plenty of taxis but try to avoid driver change-over time between 2.30 p.m. and 3.30 p.m. when they all seem to disappear. Flag fall is $1.50 and 80 cents a kilometre. If you cross the harbour bridge, the cost of the toll will be added to the fare on the clock.

The major cab companies are:

• ABC 922 2233

- Taxi Combined Services 332 8888
- Legion Cabs 20 918
- RSL Cabs 699 0144
- Radio Cabs 897 4000
- There is also a water taxi, ring Harbour Taxi on 552 2266, but ask the price first.

Rental Cars

Rental cars are quite expensive in Australia, but weekend concessions apply. The main companies have offices on William Street, close to Kings Cross. Shop around because there are quite a few smaller companies that offer better prices than the national companies. Drivers must be 21 years or over.

If you're just staying in and around Sydney it may be worth renting a second hand rental car at about half the price of conventional rental companies. Check the newspapers and Yellow Pages for details.

Driving

We drive on the left in Australia. Remember, seat belts are compulsory for all passengers, and young children *must* use approved child restraint seats.

In New South Wales, the alcohol limit for driving is 0.05 ml per litre of blood, and is strictly enforced by **random breath testing**. Drivers caught over the limit are taken for blood tests, and may find themselves in jail for the night.

Tourist Bus Tours

Bus and ferry tours are a relaxing way to see the major sights. To find the tours which best meet your needs, stock up on brochures from the bus companies. Most hotels carry a selection. Failing this, the best place for information is the NSW Travel Centre, at 19 Castlereagh Street, or the tourist information booths in Martin Place, the Pitt Street Mall, and Circular Quay. The major tour companies are Clipper Tours, Australian Pacific Tours, Sundeckers, AAT Kings, Ansett Pioneer, Murrays, Newmans and Sunliner. Check the Yellow Pages for the latest numbers. For more details on the Harbour cruises, see Part 5. For some excellent ideas on trips out of Sydney, either by tours or ones you can arrange yourself, see Part 7.

Renting Bicycles

Sydney's frenetic traffic is not the best accompaniment for a gentle cycle ride. Heavy traffic and choking fumes, combined with the narrow roads and steep hills, discourage all but the most fanatical and suicidal. Still, if you feel an uncontrollable urge coming on, take the bus to Clovelly Road, adjacent to Centennial Park, and rent a bike there (see Part 8 for details). Bike rentals are also available in Glebe, west of the city centre at Inner City Cycles, phone 660 6605.

Accommodation

This section will give you a good idea of the choices available by price and by area. Some good alternatives to the top range hotels are suggested. In Sydney's rapidly growing tourist market, we cannot guarantee that some of our facts won't quickly change. Go to the Travel Centre of New South Wales for the most up to date information. An excellent source of accommodation information is available from the NRMA (the motorist association). The NRMA accommodation guides are free to members of some sister organisations abroad, so you might like to check this out before you leave.

Choosing Where to Stay

What area you choose for your stay will depend on what you value most. Most first class hotels are in the centre of the city, so if your preference is for the top end of the market then you probably need to look no further. Even so, there are some mid-range hotels and budget hotels in the southern part of the business district and other areas.

Kings Cross has a wide variety of accommodation within a very small area, so if you're looking for accommodation, this is an easy place to try. Although it is the night club area of Sydney there are up-market hotels in the quieter parts and several backpacker hostels. The city centre can be reached in less than 10 minutes by train from the Cross.

If an early morning swim in the Pacific would start your day perfectly and you're prepared to travel a little then Bondi or Manly will give your holiday a very special Sydney flavour.

World famous Bondi is the raciest beach area in Sydney, with many coffee shops, restaurants and take-aways. Bondi has a very different atmosphere from Manly. Contrary to what you might expect, Bondi is not Sydney's version of Waikiki Beach, or even Queensland's Gold Coast. Bondi's heyday was the 1930s. Since then, Bondi has become frayed at the edges. There has been little recent investment and many of the hotels are showing the strain. There are no top range hotels in Bondi. Bus and train to city takes about 30 minutes.

Manly is more sedate and has a beautiful long beach. Again no shortage of restaurants, and take-aways. The ferry to Circular Quay and the City centre takes 30 minutes and costs about $6 return. Discount tickets help to reduce travel costs on the ferry if you'll be staying a few days. If you're in a hurry, the hydrofoil will whisk you into the city in about 15 minutes.

For a less hectic pace you might like to consider the North Shore. The municipality of North Sydney is just two stops on the train from Wynyard and the centre of the city. North Sydney is Sydney's second business district and is Australia's fourth largest office centre after Melbourne and Brisbane. North Sydney also contains the beautiful harbour suburbs of McMahons Point, Kirribilli and Neutral Bay. Picturesque Kirribilli, across the harbour from the Opera House has Sydney's biggest concentration of high-rise flats. Kirribilli has a few coffee shops, restaurants and take-aways. Travel by private or public ferry from the Quay, or take the train to Milsons Point, just one stop from Wynyard Station. Kirribilli is a good choice if you want to see Sydney as the residents do. There's not much accommodation to chose from but there is some choice in the mid and budget range.

Price Range

Sydney is not a cheap city, and accommodation is certainly no bargain. For first class hotels, you'll find the city as expensive as anywhere. The top hotels charge top prices. But excellent accommodation and perhaps better value can be found in many cheaper hotels.

Our price range is based on the latest information available in 1990. Allow some price inflation from that date. Typical price ranges for hotels in 1990 were:

High Price: over $180 for double or single.
Medium Price: $90-180 for a double, $70-90 for a single.
Budget: Under $80 for a double and under $40 for a single.
Backpackers: Around $10 for a bed in a dormitory room.

High-priced Hotels

City centre

The top range hotels are expensive by anyone's standards. Prices are $220 and up for a single and $250 and up for a double. Here are some of the choices:

Holiday Inn Menzies (20 232) on George Street at Wynyard.

Hotel Inter-Continental (230 0200) near the Botanic Gardens, which is built onto the old New South Wales Treasury Building.

Sheraton-Wentworth (230 0700) on Phillip Street.

Ramada Renaissance (259 7000) on Pitt Street.

Regent of Sydney (238 0000) near Circular Quay.
     Alternatives that are walking distance to the City centre are:

The Boulevard (357 2277) situated halfway between the centre and Kings Cross at 90 William Street.

Southern Cross Hotel (2 0987) on Goulburn and Elizabeth Streets.

Old Sydney Park Royal (2 0524) has a very pleasant location in the heart of the Rocks at 55 George Street.

Apartment hotels are another alternative in and near the city centre, usually starting around $180 in the top range. They have complete cooking facilities and one or two bedrooms. There are several places, such as *The Waldorf* (261-5355) at 57 Liverpool Street, just west of George Street, *York Apartments* (264 7747) 5 York Street, *Sydney Visitor Apartments* (290 1166) 57 York Street, *Hyde Park Plaza Hotel* (331 6933) at 38 College Street and *The Stafford* (251 6711) at 75 Harrington Street, The Rocks.

Kings Cross

The most prominently positioned top range hotel is the *Hyatt Kingsgate* (356 1234), on the corner of Victoria Street and Kings Cross Road. Other top hotels in the area include the *Sebel Town House* (358 3244) at 23 Elizabeth Bay Road which is a favourite among members of the entertainment industry and the *Gazebo*

Ramada (358 1999) at 2 Elizabeth Bay Road.

As an alternative to the top end hotels, there are a number of small boutique hotels in Potts Point, close to Kings Cross. These hotels don't offer all the facilities of the international hotels but are luxurious and have a more distinctive and individual character:

The Jackson (358 5144) at 94 Victoria Street, $80 double and $70 single.

The Kendall (357 3200) at 122 Victoria Street has been converted from two Victorian terrace houses, which retain much of their original charm. $120 double and $110 single.

Simpson's of Potts Point (356 2199) is at 8 Challis Street.

North of the harbour

In Manly, the top range hotels include the *Manly Pacific Parkroyal* (977 7666), and the *Radisson Kestrel Hotel* (977 8866) on the ocean front, at 8 South Steyne, just a few hundred metres from the Wharf. Doubles and singles from $150. *The Manly Plaza* (977 1977), at 48 Sydney Road, is 300 metres from the Manly Wharf, two blocks from the Corso, and the ocean beach. Doubles and singles from $135.

The top range hotel in North Sydney and on the edge of the centre is the *North Sydney Travelodge* (920 499), at Blue Street, just 50 metres from North Sydney station, and overlooking the harbour. Rooms from $175.

Medium-priced Accommodation

City centre

There is a broad range of medium-priced hotels in and around the centre. Here are a few suggestions:

Greetings Oxford Towers (267 8066), Waine Street, one block south of Oxford Street, at Riley Street, behind the Koala. Doubles $90 to $133, and singles $80 to $133.

The Harbour Rocks Hotel (251 8944), on Harrington Street is a new hotel in the Rocks, just one block from George Street. This beautiful and imaginatively converted warehouse and workers cottages has standard rooms at $140, and rooms with *shared facilities* at $110 double, and $80 single. Many people might not want to pay so much for shared bathroom and no TV, but for those who don't mind this is an interesting accommodation choice.

The Russell (241 3543) on George Street in the Rocks is another lovely place with character. Prices are $100 and up.

Koala Park Regis (267 6511) at Castlereagh and Park Streets, with doubles at $105 and singles at $95.

For a medium-priced apartment, try *Sydney Visitors Apartments* (290 1166). They have a range of apartments in six different locations in the city.

Kings Cross

The Clairmont Inn (358 2044) at 5 Ward Avenue is in the quieter part of the Cross and has rooms at $110.

The Crest (358 2755) just the other side of Darlinghurst Road from the Hyatt. Doubles are $145 and singles $130.

Hampton Court (357 2711) at 9 Bayswater Road is a good choice, and less expensive than The Crest. Doubles start at $98, and singles $88.

Madison's Hotel (357 1155) at 6 Ward Avenue. A new hotel, only two minutes walk from Darlinghurst Road, slightly more expensive than the Hampton Court.

Bondi

Breakers (309 3300) at 164 Campbell Parade, on the beach front, has rooms and suites (one bedroom with sitting room), from around $100, including cooking facilities. Undercover parking is available.

Greetings Bondi Beach (365 0155) at 136 Curlewis Street, has rooms with cooking facilities from $90 for doubles and $82 for singles.

The *Cosmopolitan Motor Inn* (305 311) at 152 Campbell Parade, on the beach front is a little less expensive, a bit shabby and less modern. Rooms with cooking facilities are priced between $80 and $90. Undercover parking is available.

If you want to be close to the beach but more convenient to the centre of the city, Bondi Junction might be a good solution. At the end of the Eastern Suburbs Railway line, it is just a few stops from the City, and two stops from Kings Cross. A short bus ride will take you down the hill to Bondi Beach.

Greetings Bondi Junction (389 466) at 79 Oxford Street has doubles from $85 to $103, singles from $77 to $93. *Bayline Motor Inn* (389 6134) at 3 Waverley Crescent Bondi Junction, has rooms for $60 double and $50 single. Undercover parking is available. Both are close to the centre of Bondi Junction.

North of the harbour

Northside Gardens (922 1311) at 54 McLaren Street is good value in the medium-price range. On the north side of the business district, and about 10-15 minutes walk from the station this may be a good choice if you have a car, and want to explore areas beyond the city centre. Rooms range from $80 to $135. Suites which include cooking facilities are $170 to $190.

Kirribilli Lodge (923 1822) at 33 Fitzroy Street, has doubles at $92 to $99, and singles at $80 to $85. Suites with cooking facilities are $107 to $160.

Kirribilli Guest House (922 3134) at 12 Parkes Street in Kirribilli, has doubles at $75, and singles $50, including breakfast. There is also some cheaper and longer stay accommodation in Kirribilli, try the *Elite* (929 6365), at 133 Carabella Street.

Manly

Manly has some good alternatives for the budget-minded. See also the motels listed in the budget section.

Manly Surfside (977 2299) has one, two and three-bedroom holiday apartments available, at between $100 and $150 a night.

Manly Paradise Beach Plaza Apartments (977 5799) has one and two bedroom units from $110 to $145.

Parramatta

Although it is a half hour journey by bus or train the *Parkroyal* in Parramatta (689 3333), 30 Phillip Street is a good alternative for high quality accommodation in this price range. They often have special family rates.

Budget Accommodation

City centre

There are a few choices in budget accommodation in the centre of the city, although you may find some of it a little worse for wear.

Cron-Lodge Motel (331 2433) 289 Crown Street, Surry Hills. Double rooms are $64 and singles $55. All have private facilities, and lock-up parking is available. It is three short blocks south of Oxford Street, so convenient for the Pioneer-Greyhound bus terminal, mid-way between the city centre and Kings Cross.

Great Southern (211 4311) 717 George Street, 500 metres south of the Central Railway station. Doubles $46 and singles $32.

YWCA (264 2451), 5-11 Wentworth Avenue in Darlinghurst, just off Liverpool Street by Hyde Park. This large modern YWCA is open to women, couples, families and groups. Single men are excluded. Price starts around $35 for single with shared bathroom and $60 with private bathroom. Twin rooms are $55 with shared bath and $70 with private bathroom. They have triples and also dormitory rooms for four people at $15 per person. It is handy to the Museum railway station, buses, and the central area.

Another alternative in the budget end of the market if you want to stay in the central area are the rooms above the pubs. Most of them cannot be described as cosy and many are need of updating. Bathroom facilities are generally shared. The other drawback is that if you want an early night the noise in the bar doesn't stop until 11 p.m. The pluses are that they are centrally located and inexpensive at about $40 for a single and about $50 for a double. A couple of possibilities are:

The Harbour View Hotel (27 3292) at 18 Lower Fort Street, The Rocks, *The Criterion* (264 3093) Park Street and Pitt Street (sadly this is threatened with redevelopment and may not survive much longer). And tucked in behind the Rocks area is *The Palisade* (27 2272) at 35 Betttington, Millers Point.

Kings Cross

The budget accommodation is a mixed bag. As there are so many hotels so closely situated, it is easy to check out a few and find the one which best suits your needs. Most are small so consequently any one may be full when you want it.

Carnarvon Lodge (358 6611) at 16 Ward Avenue, just along from Madison's. Double rooms here range from $70 to $85, and singles are $60.

The Barclay (358 3122) at 17 Bayswater Road, which is just across from *The Crest* and less that 100 metres from the station. Rooms here range from $55 to $90 for doubles, and $ 45 to $90 for singles.

Holiday Lodge (356 3955) at 55 Macleay Street is a small but very pleasant hotel with friendly management. Rooms here are $70 for a double and $60 for a single. Parking is a problem, booking ahead is essential if you need it.

Kirketon Hotel (360 4333) at 229 Darlinghurst Road, doubles at $58 and singles at $38.

Network Travellers Hotel (358 3259) at 6 Orwell Street is only a few blocks from the station, and offers rooms from $35. This is designed to appeal to well-healed backpackers, so don't expect the standard hotel presentation. It is a good opportunity to mix with other travellers, and for couples it is a good deal.

For an apartment hotel try *Kings Cross Holiday Apartments* (360 4610), at 169 William Street, Darlinghurst. A studio costs $70.

Kings Cross has many backpackers hostels and hotels. See the backpackers section for details.

Bondi

Alice Motel (30 5231) at 30 Fletcher Street, Bondi Beach, has doubles at $60-$70, and singles $50-$60. All rooms have TV and refrigerator. Free undercover parking is available. A little bit away from the action, but a reasonable price for the area.

Manly

Manly has a good choice of budget accommodation. Here are a few examples:

Manly Surfside (977 2299) at 102 North Steyne, with doubles between $70 and $80, and singles $65 to $75, and balconies each with a view of the sea.

Manly Paradise Motel (977 5799) at 54 North Steyne is also the ocean front, from $69 single and double.

The Steyne (977 4977) a licensed hotel on the Corso, Manly's main shopping street. Rooms here are $52 to $68 for doubles, and $26 to $34.

Periwinkle Guest House (977 4668) at 18 East Esplanade offers doubles at $66 and singles at $60. A light breakfast is included. The Periwinkle is on the harbour side of Manly, just a short walk from the Manly Wharf. The ocean beach is just two blocks away. Rooms include refrigerators.

Other places outside the central city

The Glebe, close to Sydney University is an interesting place to stay, and is only a short 431 bus ride from Central Station.

Rooftop Motel (660 7777) at 146 Glebe Point Road, has rooms for $70.

Hereford Lodge (660 5274) at 51 Hereford Street has basic motel-style rooms without TV or phones but they are clean and very reasonably priced at $45 a single and $55 a double.

Some of the student residences offer accommodation during university vacation periods. For example, *The Women's College* at the University of Sydney (51 1195) offers rooms for students and youth hostel members at $24 for a single and $42 for a double, bed and breakfast. Non-student rates are about 25% higher.

Bed-and-breakfast in people's homes can be booked through the *At Home Down Under* scheme. Cost per person starts at $35 for a single and $42 for a twin with economy, quality, and superior to choose from. Phone 960 4481 for details and bookings.

Backpackers and Youth Hostels

At any one time there are thousands of backpackers in Sydney, but only recently has there been much interest in supplying them with accommodation. Now there should be no difficulty finding dormitory and low cost accommodation, but take some care in checking the options. Expect to find some rip-off artists around. As always travellers are the best guide to good value. As hostels can change in character very quickly, we have listed some from the wide range available. There are new ones being added all the time. See also the listings under budget accommodation above.

YHA now has six hostels in Sydney. Check YHA handbook for full details. Accommodation is for YHA members only. All hostels are open all day and there is no night-time curfew. Office hours vary but all of the hostels are open from 7 a.m. to 10 p.m.

Three hostels in the Glebe area, close to Sydney University and a quick bus ride from Central railway station provide a mixture of bunkrooms and twin and share rooms:

Hereford Lodge (660 5274) at 51 Hereford Street has some luxurious extras such as a large cafeteria open for breakfast and dinner and a roof top swimming pool and sauna. Prices are between $12 and $16 per person for twin rooms or shared rooms, with children half-price. Motel-style rooms also available.

Forest Lodge (692 0747) 28 Ross Street are $12 per person and has bunkrooms only.

Glebe Point Hostel (692 8481) 262 Glebe Point Road.

Dulwich Hill (569 0272) hostel at 407 Marrickville Road, Dulwich Hill, has twin and family rooms and also accepts groups. Get there by 426 bus from Railway Square, or train on the Bankstown line to Dulwich Hill station.

Pittwater Hostel (99 2196) is situated in the bush about 40 kilometres north of the city, and is reached by ferry to Manly, 155 bus to Church Point and then ferry to Halls Wharf. You must phone ahead.

For commercial backpackers accommodation, Kings Cross is the main location. These are concentrated around Darlinghurst Road and Victoria Street and are convenient to the Kings Cross railway station. The Kingsford-Smith Transport minibus from the airport will drop you directly at many of these hostels.

Kings Cross

It pays to spend a few minutes looking around as standards can vary substantially (cockroaches may accompany in-room cooking facilities). *The (Original) Backpackers* (356 3232) at 162 Victoria Street was the first of the backpacker-style accommodation and has maintained a good reputation, with long office hours and helpful staff. Prices for a twin room per person start at $12 and dorms at $10. Recommended by travellers as secure, an especially important feature at the Cross, is the *Jolly Swagman*. At last count there were three separate buildings, all in the same area. These are 14 Springfield Mall (358 3330), 16 Orwell Street (358 6600), and 144 Victoria Street (358 6400). The offices are staffed only part time. Rooms are $13 per person twin, and $10 share, and rooms have fridge, stove and cooking utensils. Another chain, slightly more expensive, is the *Backpackers Headquarters*, with three locations at the Cross, and two others close by. For the Cross try 79 Bayswater Road (331 2525), 59 Bayswater Road (332 3284), and 6 Orwell Street (358

2185). *Kanga House* on Victoria Street is clean but the rooms are tiny. Prices are around $11 for a bed in a shared room. All of these are within a few minutes of the station.

City centre

Backpackers (698 8839) at 243 Cleveland Street, just behind Central Station, with shared rooms from $8.

7 Elizabeth Street (223 3529) three blocks east of George Street, at Wynyard Station, for $14 a night.

Bondi Beach

The Lamrock Hostel (365 0221) 7 Lamrock Avenue, Bondi Beach, just one block from the beach, off Campbell Parade (train to Bondi Junction and then bus to the Beach).

Manly district

Manly Lodge (977 8655) at 22 Victoria Parade in Manly has backpackers accommodation, but depending on demand may be available only on a weekly basis.

Avalon Beach Hostel (918 9709) at 59 Avalon Parade, Avalon has beds at $10 a night. Avalon is well north of Manly close to Palm Beach; reach it by the Palm Beach bus from Wynyard or Manly.

PART II
History

History and Architecture

Approaching the entrance to Sydney harbour by ship, the visitor is struck by the majesty of the natural landscape. North Head and South Head mark the entrance to one of the world's great harbours, and provide the most spectacular entry point of any Australian city. This gateway is much as it was in 1788 when Europeans first sailed between the Heads, intent on establishing a settlement in what they regarded as a virtually uninhabited continent.

Today, the visitor may notice dwellings crouched along the cliffs of the eastern suburbs, but the cliffs themselves remain much as they have been for thousands of years. Inside the harbour, many visitors are surprised to find that much of the foreshore remains unalienated, especially on the northern side.

Most overseas visitors who come to Sydney arrive not by ship, but by aeroplane. Because Mascot Airport is only ten kilometres south-east of the city, most air passengers are greeted with a superb overview of the ocean, the harbour, the city centre, and a great sprawl of detached houses to the north, south and west. Even in the flatter parts of the city the skyline is now dotted by regional centres, many boasting a number of high-rise office towers.

Sydney was never officially named; it derived its name from Sydney Cove, where settlement was first established. That name, given by Governor Arthur Phillip, honoured Thomas Townshend, 1st Viscount Sydney, then Home Secretary in the British cabinet.

Penal Settlement and the Aborigines

When the first party of convicts arrived with Captain Arthur Phillip in January 1788, they landed at Botany Bay. Finding it relatively shallow and unprotected, lacking good water and fertile land, they moved to the deeper waters of Port Jackson. Phillip's party was much struck, as Captain Cook and the botanist Sir Joseph Banks had been before them, by the terrain and the vegetation. The judge advocate of the colony, David Collins, wrote:

> The spot we've chosen ... was at the head of the cove, near the run of fresh water [the Tank Stream] which stole silently along through a very thick wood, the stillness of which had then for the first time since creation been interrupted by the rude sound of the labourer's axe, and the downfall of its ancient inhabitants. (*An Account of the English Colony in NSW*, London, 1798)

In 1788 there were probably 3000 Aborigines living around Port Jackson. Fifty years later there were fewer than 300. Aboriginal settlements around Sydney Cove were quickly obliterated and the "forests", as the English called them, soon fell to the labourer's axe.

Nonetheless, extensive evidence of Aboriginal occupation remains in many parts of the Sydney region. The Europeans did not understand how the Aborigines had used the land. They used fire to force animals out into the open and to create patches of green regrowth which would attract kangaroos. On the Hawkesbury sandstone which surrounds the Cumberland Plain—the area between the coast and the Hawkesbury/Nepean Rivers—they drew familiar objects: wallabies, fish and eels, boomerangs and spears. Each of the three major tribes in the area—the Dahrug, Dharawal and Kuring-gai—had access to sandstone outcrops suitable for painting and rock engraving. Thousands of engraving sites still remain within 100 kilometres of Port Jackson. The engraving sites most readily accessible to the visitor are those within the Ku-ring-gai Chase and Royal National Parks to the north and south of the city. Until recently the National Parks and Wildlife Service has been loathe to reveal the location of many of these sites because in an Australia where Aboriginal culture went unappreciated, vandalism was common.

The early penal settlement consisted largely of tents but they were gradually replaced by wooden huts. By March 1788 the convicts were employed in the manufacture of bricks with clay taken from what today is known as the Haymarket. In May 1788 Governor Phillip directed that work begin on the first major permanent structure to be erected in Australia, a two storey brick dwelling which became the first government house. The house boasted the colony's only staircase, a "source of wonderment to Phillip's Aboriginal visitors, who were astonished to find people walking about above their heads". It had a commanding view of Sydney Cove and rapidly became the centre of administration as well as the Governor's residence.

Macquarie's Sydney: Seat of Government

Twenty-two years after settlement, in February 1810, when he had been in Sydney for two months, Governor Macquarie wrote of a fledgling settlement in rapid decay:

> On taking command I found the Public Buildings of every description in a state of shameful dilapidation, or of rapid decay. The streets of Sydney were almost impassable, and the principal roads and bridges were, if possible, in a still more dangerous and neglected state.

For eleven years Macquarie did his best to turn Sydney into a seat of government of which its inhabitants could be proud. He issued regulations for the widening and straightening of streets and he pursued a vigorous policy of erecting public buildings, some of which survive. In 1810 work began on a new general hospital, popularly known as the "Rum Hospital". In October 1810 the new Governor named the street on which the hospital stood "Macquarie Street" which he described as "the easternmost street in the town, extending in a southerly direction from the Government Domain to Hyde Park". The park, which he also named, was nothing more than cleared bushland which served as a race track. (Berzins 1988; Ritchie 1986)

One wing of the Rum Hospital came to serve as Parliament House and another as the Royal Mint. Other buildings that have survived include Francis Greenway's convict barracks (Hyde Park Barracks 1817), such an elegant Georgian structure that it almost belied its purpose; some rather extravagant stables for Government House (1821) that later became the State Conservatorium of Music; St James Church (1820-24), and the Supreme Court (1820-28). The Barracks, St James, and the Supreme Court were designed to harmonise with each other. All were built of soft red brick with sandstone trim (Kerr 1981). The axial arrangement of St James and the Barracks across the head of Hyde Park were planned, with the Hospital/Mint and the Supreme Court, to make a civic ensemble, now embellished by street closures to form Queens Square, one of the few consciously planned urban spaces in the city. All these buildings are readily accessible from St James Station. The Barracks and the Mint are open to the public as house museums and regular tours are conducted of Parliament House.

Churches, public buildings and military installations of the time were usually built in stone, and this substantially contributed towards Sydney assuming a more urban appearance. Fort Denison (1840-57), constructed for harbour defence, and Bare Island Fort at La Perouse (1880s), built to counter a feared Russian invasion, are two well-known sentinels.

One group of commercial buildings remains from this era, namely the bonded stores to be found in the Rocks, including the Metcalfe Bond (1821-4) and the Argyle (1828). Now the site of one of Sydney's leading tourist attractions, the Argyle Arts Centre, it was the first enterprise to capitalise successfully on the new-found interest in Australia's early history that developed in the 1960s. The Rocks now bills itself as the "birthplace of Australia", which is not strictly accurate, but as it has a larger concentration of pre-1830 stone buildings than any other city except Hobart, the claim has not been widely disputed.

Harbour and Hinterland—Town to City

Although founded as a penal settlement, Sydney's port function developed early. Shipping was the only form of communication with the mother country and for most of the 19th century the main form of communication with most other parts of the continent.

From the 1830s, Sydney's population began to expand rapidly, reflecting the city's growing role as a port for the export of produce from the hinterland. The population increased from 11 000 in 1828 to 30 000 in 1841 and this was reflected in an expansion of the main settlement beyond the one square mile that had been intensively occupied by 1820. In the 1830s, land was subdivided at Surry Hills, Woolloomooloo, Pyrmont, Paddington, Redfern, Glebe and Newtown and villages were also established at Ashfield and Burwood. This period also saw the construction of more elaborate buildings such as the elegant **Elizabeth Bay House**, designed by John Verge and built between 1832 and 1837. It is a classical, colonial version of late Georgian architecture, with a fine cantilevered stone eliptical staircase. For much of the 20th century, its rooms were carved up for use as a boarding house, but in the 1970s it was restored to its former glory and opened as a house museum. Government House, built between 1837 and 1845, occupies a commanding site, overlooking Circular Quay. By the 1840s Sydney was well on the way to shedding its appearance as an urban gaol.

Sydney: the Nation's Major Port

Exports of wool began in substantial quantities in the 1820s and in the 1830s whaling became an important industry. Until the 1830s almost all manufactured goods were imported but a shipbuilding industry soon developed. A network of wharves had to be constructed to cater for the growth of imports and exports.

The first wharves were primitive quays built on the western

Circular Quay around 1870 ... (*National Library*)

side of Sydney Cove. Between 1900 and the early 1920s, the government-owned Sydney Harbour Trust rebuilt wharves, using finger wharf technology, in Woolloomooloo Bay, Walsh Bay, and Darling Harbour. Of these the Walsh Bay wharves and the Woolloomooloo finger wharf are the best remaining examples of wooden finger wharves anywhere in the world (Holt and Spearritt 1987).

As free settlers came to outnumber convicts, demands for self-government at both the local and colonial level became more vociferous. A Sydney Municipal Council was established in 1842 with a wide range of functions, including public health and sanitation, street maintenance and lighting, markets, water supply and partial control of the police. In the next 50 years most of these functions were passed on to colonial departments or statutory authorities.

Sydney's growth was helped by the fact that until the opening of Melbourne's Williamstown dock in 1874, it had the only two dry docks in the southern hemisphere: Mort's Dock at Balmain opened in 1855, and the Cockatoo Dock opened in 1858. By

... and as a busy ferry terminal early this century

1886 Mort's Dock shipbuilding and repairing operation occupied a site in Balmain: Sydney was well on the way to becoming a great port city. Throughout the rest of the 19th century it maintained this lead, effectively stifling the development of possible rivals such as Newcastle.

Most wool continued to come through the port of Sydney, even if that involved transhipment from steamers from other New South Wales ports. Many of the warehouses built for wool and other rural produce are still to be seen on the shores of Darling Harbour today. In the last twenty years, tonnage through the port of Sydney has almost been equalled by the new container port at Port Botany (in 1986-87, 18 million tonnes to 14.4 million tonnes). However, Port Jackson remains the major port for general cargo and the maintenance of vessels.

Shipping is synonymous with the life of central Sydney and has become a major tourist attraction. Ferry services from Circular Quay, which languished after the opening of the Harbour Bridge in 1932, are re-gaining patronage every year. Some of the historic finger wharves have been converted to new uses—most notably the Sydney Theatre Company and its superbly sited restaurant "The Wharf", at the tip of one of the Walsh Bay wharves. In 1987 the state government announced that the Woolloomooloo finger wharf would be preserved. Since then its fate has been in limbo, attacked by some politicians and commercial developers as ugly, while others have argued that it is an integral part of Sydney's heritage.

Servicing the Metropolis: Water, Gas and Electricity

By the 1820s the Tank Stream was an inadequate and somewhat polluted source of water, and in 1827 work started on a tunnel known as Busby's Bore to bring water from the Lachlan Swamps (now Centennial Park) to the city. Completed in 1837, Busby's Bore was superseded in 1858 by a pumping scheme drawing water from the Botany swamps. The Victorian pumping stations can still be seen from the airport freeway. Responsibility for water supply was taken over in 1888 by a Metropolitan Water and Sewerage Board and ceased to be a concern of the City Council.

The city first experienced gas light in 1841 and gas continued to be used for street lighting until the 1890s when it was gradually replaced by electricity. Some original gas lights remain in Argyle Place, the Rocks—a reminder that gas was the dominant mode of lighting for more than half a century. Huge gasworks and gas tanks appeared in suburbs as far flung as Manly to the north and Mortlake to the west of the city. All the gas tanks have now been demolished, with the move to natural gas.

By 1890 several hotels, theatres, emporia and newspaper offices were lit by electricity. Most of these early installations were powered by their own steam engines. The City Council had lost its responsibility for the water supply and had never had it for gas, which was in the hands of a private firm, AGL. Thus, the Council was keen to seize the initiative in electricity and in 1902 began construction of a power station at Pyrmont, which, in a much modified form, continued to function until the early 1980s.

Some of Sydney's grandest power stations, such as Bunnerong overlooking Port Botany, have now been partially demolished. But others survive, including Pyrmont, White Bay and Balmain. The Ultimo power station is now the site of the Powerhouse Museum. After World War II, power stations were no longer built near the centres they were supplying, but were erected instead near the supply of energy, so most of Sydney's power now comes from power stations near the Newcastle and Lithgow coal fields.

Transport

Sydney has been, and always will be, captive to its topography—which makes it at once beautiful and geographically complicated. The harbour made transportation extremely difficult because the city was almost a series of peninsulas, and ferry traffic was restricted to Port Jackson, Middle Harbour and the Parramatta River.

Until the 1880s, the railway line from Redfern to Parramatta, opened in 1855, remained the only line in the Sydney region. In a city with everything within walking distance, the bulk of the population—who could not afford to maintain or house a horse—congregated in and around the city centre to remain within easy access of their workplaces. Horse-drawn omnibuses provided an alternative means of getting about for those who could afford the fare.

In the 1880s, with the establishment of a tramway system and the construction of more rail lines, this pattern began to change for it became possible to live some distance from one's work. The near-town settlements, served by a network of trams, started to spread into a network of suburbs. But the system remained centred on the city proper. Trams were first drawn by horses, then powered by steam or cable, and electrified from the late 1890s.

The tramway and railway boom is still much in evidence in central Sydney, despite the fact that trams were withdrawn in the 1950s and early 1960s. Tram lines can still be found not far below the surface of many a city and suburban street, and small remnants of tram shelters—characteristically in wood with tile roofs—are still to be seen in and about the city. But the biggest tramway installation in central Sydney—the tram terminus at Bennelong Point (designed by Vernon in 1902)—was demolished in the 1950s to make way for the Opera House.

The symbolic centre of Sydney's and New South Wales' railway system remains the gigantic Central Station opened in 1906. The clocktower at Central was one of the tallest structures of its time, and even today it retains a dominant position on the Sydney skyline, a vital part of the "sense of Sydney". Before the opening of Central the main Sydney terminus had been a third of a kilometre westward. The gothic Mortuary Station at Regent Street, dating from 1869, is the most notable survivor from this earlier era. Funeral parties set out from the Mortuary Station for Rookwood Cemetery, a 15 kilometre train journey. The railway station is still there, but it is not in use.

Australia's first underground railway was opened in 1926, when St James and Museum stations, designed to combine the best of the London and Paris undergrounds, were unveiled to the public. Even more spectacular was the opening in March 1932 of the Sydney Harbour Bridge, together with Wynyard and Town Hall Stations, for the completion of the link across the harbour. About that occasion a Sydney Morning Herald jour-

Construction of the Sydney Harbour Bridge

nalist, who had hired a light plane, reported a "city of deserted suburbs":

> They are the streets as of a city of the dead. No pedestrian, no motorist is seen. Streets, and houses, and backyards—devoid of life ... below to the left the lifelessness ceases. Hyde Park and the streets that run towards it are choked with people ... suddenly one realises that to the west is the focus of all this activity—the bridge, curved like a bow half-drawn ... a colossus in its mighty stride.

When work on the bridge began, it was to be the longest, widest, and heaviest arch bridge in the world. But the Bayonne Bridge over the Kill Van Kull River between New York and New Jersey, opened in 1931, was deliberately extended to have a span 60 centimetres longer. However, the Sydney Harbour Bridge remains wider and heavier than any other arch bridge in the world.

The opening ceremony by radical Labor Premier Jack Lang—who had antagonised the banks and the insurance companies with his left-wing policies during the Depression—was disrupted by Francis de Groot, a member of the New Guard, a fascist organisation. Disguised as a military horseman he rode up and slashed the opening ribbon in advance of the Premier. Newspapers throughout the world published a spectacular photograph of de Groot's action.

The underground loop was not completed until 1956, when the Circular Quay Station and Cahill Expressway were opened too. The Edwardian wharves at the Quay were rebuilt in a more modern style from the late 1940s and again for the bicentennial year, 1988. They are now completely open to the water on three sides, affording superb views of the harbour.

Terraces, Houses and Flats

In 1829 Burford wrote in his *Description of the Town of Sydney*:

> The houses are in general substantially built of free stone, or brick plastered, seldom more than two or three storeys in height, with verandahs to at at least one storey; the roofs are covered with shingles of iron bark, or cedar. The original buildings, which were of soft wood of the cabbage palm, have long since decayed; indeed, but very few of the earliest brick buildings remain. Near the harbour, where the ground is valuable, the buildings are contiguous; but generally speaking, the better sort of houses are detached, having a small enclosure in front, with a neat geranium hedge.

The emerging classic form of Australia suburbia, with detached houses and ample gardens, was thus early remarked upon. The range of housing available in Sydney during the mid-19th century is well-illustrated in Bernard and Kate Smith's book *The Architectural Character of Glebe*, Sydney (1973). Most housing built in Glebe between the 1840s and the 1860s can best be described as post-Regency:

> A style of blocky simplicity in mass and contour, built in stone or rendered brick, possessing clear lines and classical proportions in the treatment of windows and doors, a dominating rectangularity and broad eaves lines.

Picturesque Gothic also had its heyday in the 1860s, but by the 1870s the Italianate style had come into its own. Terrace houses took on a Renaissance palace air, as did detached houses. Decorative plasterwork and fancy cast-iron lacework were common features, along with parapets and bay windows. Towers and belvederes are often to be found on larger houses. By the 1890s the Federation Style was having its impact in Glebe, but plentiful examples are also to be found in more distant middle-class suburbs such as the developments around Centennial Park, Mosman, and along the Hume Highway through Ashfield. The pre-eminent characteristics of the Federation Style, sometimes called "Queen Anne", are of terracotta roof tiles and exposed bricks of a deeper red.

> Carved, fretted or turned wood, usually painted white, replaces cast iron balconies and verandahs... The roof is broken into many divisions and these often reflect the disposition of the rooms beneath. Flying gables are used in profusion, with shingles, ornamental half-timbering and rough-cast work appearing in beneath gables and eaves... Windows take many forms, straight, segmented or round-headed, but they are usually narrower in proportion than Italianate windows.

Sydney probably has more examples of Federation housing than any other Australian city, because although it shared in the boom of the 1880s, continued growth in Sydney's rural hinterland allowed it to avoid the spectacular collapse of building activity that occurred in "Marvellous Melbourne". Moreover, the railway building boom of the 1880s continued into the following decade. Suburbs radiated from the city along the railway lines.

By 1895 most of Sydney's present suburban rail system was completed. In contrast, the tramway system continued to expand rapidly to service areas between rail lines right up to the outbreak of World War I.

The most common form of housing in Sydney at the turn of the century, as in Melbourne, remained the terrace house. In the inner-city industrial suburbs, these were usually narrow single-storey residences, often only 12 feet in width. Most were

Early colonial house

1840s terraces

Gothic sandstone, c 1840

Federation style

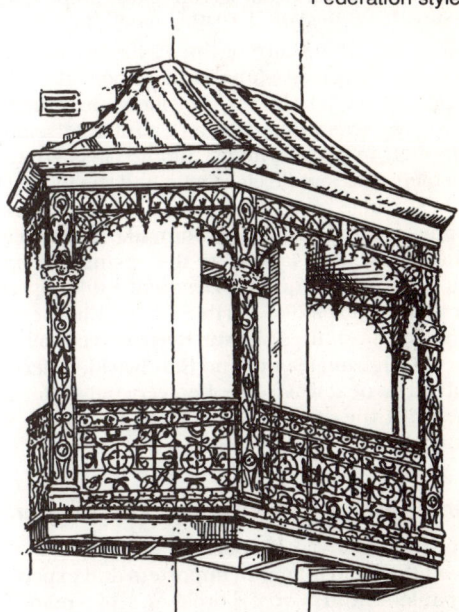

Lacey balconey, Paddington terrace, c 1890

Californian bungalow

owned by individual or corporate landlords and many had inadequate bathing facilities. Hundreds of these terraces were demolished in the slum reform movement of the 1930s and 1950s but the remainder have since been rehabilitated, first by post-war European migrants and more recently by a new generation of the middle class, who rediscovered inner-city living. In a city where the great bulk of cultural and educational facilities are still within five kilometres of the city centre, the inner suburbs have much to recommend them.

The most popular style of housing in the 1920s was the California bungalow, of which many examples are still to be found in the then rapidly expanding suburbs like Ashfield, Canterbury, Drummoyne, Lane Cove, Randwick, and Waverley. In his *Australia's Home* (1961), Robin Boyd sums up the style thus:

> It began with a rugged, simple homespun look, with heavy beams and rough textures, but soon became conventionalised. In this period the chimney was moved to the front wall as a feature of the facade. Gables were lower, and massive pylons supported a flat roof to the porch. Cement tile-roofs, shingled gable-ends, clinker-brick chimney, rough-cast walls and green paint were characteristics. A favourite name was "Green Gables" (p. 78).

Australian architecture of the 1930s became rather more adventurous than at any time since the development of indigenous styles like the wide-verandahed house. Such diverse overseas influences as Spanish Mission and Bauhaus made their impact in many middle-class suburbs, because it was chiefly the middle class who had the money to experiment. The best range of 1930s housing can be seen in some of the localities around Middle Harbour, like Seaforth and Balgowlah.

The 1930s also saw, after the easing of Depression conditions, a great wave of Art Deco flat building, concentrated in near-city suburbs like North Sydney, Elizabeth Bay, and Darlinghurst, and in suburbs well-served by rail (including Strathfield), tram (Bondi and Randwick), or ferry (most notably Manly). The next wave of—often very ugly—flat building did not come until the 1960s. Since then, flats have spread to most suburbs.

Office Blocks, Emporia and the Skyscraper Age

Central Sydney was the centre of the major retailers, and it housed the head offices of importers and exporters, of shipping companies and wool brokers, of manufacturers and wholesalers. The city's dominance over the commercial life of the metropolis reached its apex in the 1920s and 1930s, when all the large emporia were still to be found in or near the city centre. Grace Bros Broadway (1896+), Anthony Hordern's Brickfield Hill Store (1904), Mark Foys (1910+), David Jones (George Street 1887-1938, Elizabeth Street 1926, Market Street 1938), and Farmers (1910) were all within walking distance of a railway station. This was also true of the great office blocks of the era, the T&G building (1930) near Museum Station; the AWA Tower (1937) almost on top of Wynyard Station; the new banking and insurance buildings overlooking Martin Place; and the brokers, merchants and bankers' offices between Bridge and Hunter Street.

The most elegant of the department stores is David Jones' Elizabeth Street store, whose opening followed soon after that of St James Station opposite. It still retains many of its 1920s

Central Sydney before the abolition of the building height limit ...

features. Among the most notable office blocks are the Grace Building (1928) on the corner of King, York and Clarence Streets; the Commonwealth Bank (1928) in Martin Place; BMA House (1929) in Macquarie Street; Science House (1931) in Gloucester Street; and City Mutual Life (1936) on the corner of Hunter and Bligh Streets.

These office blocks—all rising to the 150-foot height limit—represent a range of architectural styles, from the striking classical columns of the Commonwealth Bank with its Beaux Arts banking chamber and the Neo-gothic of the Grace Building to the Art Deco adventurism of the City Mutual Life building.

In the interwar years one had to "go to town"—which then had only one meaning—for all but the limited goods and services that could be provided by suburban shopping centres. Certainly the suburbs had banks, but their services were limited; they had shops, but nothing to rival the city emporia; they had general practitioners and dentists, but no specialists; and they had food shops, but nothing to rival the offerings at Haymarket.

Most people bought their regular food supplies at suburban centres and sought some basic services there. But "town" was the place to be, especially in an era when most extended families would have at least one family member working in town.

Central Sydney's dominance of retailing did not seriously begin to decline until the 1950s, when more and more families were able to afford a car. Huge new car-based shopping centres were built in the suburbs. By 1990 these centres—which were uniformly ugly but at least had the merit that once inside the complex they were safe for pedestrians—accounted for over 50 per cent of all retail sales.

In April 1957 the Height of Buildings Act (1912) was amended. For 45 years the Act had prevented buildings rising more than 150 feet above ground level. Introduced because of concerns about adequate sunlight and fire safety the Act effectively kept central Sydney free of the "Manhattan syndrome" for the first half of the 20th century. The AMP insurance company was the first to take advantage of this new opportunity. By 1961 they demolished Goldsborough Mort's impressive wool stores to become the proud owners of Australia's tallest building, rising 26 storeys above Circular Quay. In 1967 the state government laid claim to the title, with a 38-floor office tower on Macquarie Street. The following year

... and before its conquest by the motor car

Lend Lease opened Australia Square Tower, a 50-floor office block designed by Harry Seidler. The AMP replied with another skyscraper, right behind its first, but lost the "tallest" title two years later when rival insurance company MLC opened a 68 storey, Seidler designed block overlooking Martin Place.

In the 1970s the race for the tallest building waned, partly as a result of the property crash in the middle of that decade. But a few years later the AMP got into the act again, funding Centrepoint (now marketed as "Sydney Tower"), Australia's tallest structure—but not its tallest building, that title having gone to Melbourne.

Heritage, Development and Demolition

Attitudes to "old buildings" have changed dramatically over the last 50 years. Even the meaning of the term "old building" has altered. In the 1930s the only buildings that were regarded as old were from the convict era and even historians and architects with a penchant for the past showed little interest in anything other than colonial Georgian. During the debate in the 1930s about whether the Hyde Park Barracks should be demolished to make way for modern law courts, one correspondent to the *Herald* wrote (Sydney Morning Herald, 17 October 1938):

> There are no old buildings in Sydney which could be preserved on account of their beauty or even for their supposed Macquarie design. All these old barns are merely out of date and can never be antiquities. Certain old buildings at Circular Quay are certainly reminiscent of olden days, but they are useless monstrosities, just the same. Words fail to describe that shocking old Barracks at the top of Macquarie Street which is a replica of many such buildings in England (clock and all) and which has nothing colonial about it. This place, made of brick and iron, will shortly become a menace and should be considered accordingly.

Although the Barracks drew a few supporters, the building only survived because the government of the day did not want to outlay capital on new courts. World War II provided a further reprieve. Much of pre-1850 Sydney had already been demolished or was about to go under the demolishers' hammer. The convict built Commissariat Building of 1812 (also known as the Old Taxation Buildings) was demolished in 1939 to make way for new headquarters for the Maritime Services Board. The demolition occasioned the practical comment in the press that 200 000 slates had to be removed from the roof. Replanning Committees for Macquarie Street and Circular Quay recommended the demolition of most 19th-century structures, including Sydney Hospital and the Bond Stores on the east side of Circular Quay. The Stores were to be replaced with modern blocks of flats. They were not, in fact, demolished until the 1950s, when they were replaced by modernist office blocks (Spearritt 1985).

In 1955 the architect and scholar Morton Herman undertook to write a book about Victorian Sydney. To that end he viewed nearly 20 000 buildings, inspected 6000 and photographed 700. The result is reported in his book *The Architecture of Victorian Sydney*, published in 1956. In his opening chapter he wrote (1956, p.7):

> In the middle of the twentieth century Sydney is still a Victorian city. Here and there in the commercial centre appears a towering building of steel and glass somewhat like a gold tooth in a discoloured Victorian mouth. A few tattered early colonial buildings, like pale ghosts from a vanished age, still have useful lives; but elsewhere the town halls, the cathedrals, many of the hospitals, houses, and most of the office buildings, came into being while Queen Victoria lived.

Herman pointed out that at "eye-level" Sydney had modern shops running in "unbroken ranks" below the awnings. But "above the visual and psychological break of the ugly awnings nearly all is Victorian". He noted that the advent of cantilever awnings in Edwardian times had seen the demise of most remaining colonial verandahs and had split Sydney into two levels. "It is this two-level city which deceives citizens in the middle of the twentieth century into believing that they inhabit a modern-built city" (1956, pp.7, 188).

In the office booms of the 1960s, 70s and 80s many of Herman's hated awnings disappeared to be replaced by sunless setbacks where citizens sheltered from the wind and rain in mean foyers. But Herman's split level Sydney can still be seen in sections of Pitt Street, Liverpool Street, Bathurst Street, Hay Street, Campbell Street, and most notably in Oxford Street where the Heritage Council has encouraged individual property owners to adopt a late Victorian colour scheme for their upper storeys.

Herman wrote at a time when much of Victorian Sydney was being destroyed. His front endpaper, showing the Sydney

Central Business District, indicated "buildings demolished while this book was being written". Almost all these buildings were in a rectangle bounded by Circular Quay, Macquarie Street, York Street and William Street. Very few demolitions were underway between York Street and Darling Harbour, while benign neglect by the Maritime Services Board preserved the Rocks and Millers Point. But in the next two decades approximately half of the buildings that Herman identified were demolished. Both grand and modest Victorian structures were destroyed; shops and houses, warehouses and woolstores, offices and all banks disappeared.

Here and there some modest buildings survive, but the main reminders of the Victorian era are the grand public buildings of the time: the General Post Office, the Town Hall, the Queen Victoria Building, and the head offices of government departments clustered around Bridge Street. Still, the City has also gained two notable cultural facilities in the last two decades: one an ornament—the Sydney Opera House—and the other a stark concession to mass culture—the Sydney Entertainment Centre.

Religion, Education and Culture

Sydney began as an Anglican settlement. The first two Catholic priests, both Irishmen, were not admitted until 1820 and they did not gain permission to build a chapel until the following year. The first Presbyterian minister, John Dunmore Lang, did not arrive until 1823. The Anglican Church lost its monopoly of public religious observance in Governor Bourke's Church Acts of 1836, which enabled the colonial government to grant land more equitably to the major Christian denominations.

A rash of church building followed. The Anglicans already had their elegant St James Church (1819-24) at the top of King Street, but in 1837 they embarked on a Cathedral, St Andrew's, on George Street. The Catholics did not embark on their present Cathedral, St Mary's, until the mid1860s. But even by 1928 it was only 90 per cent complete, and its spires have still not been added (O'Farrell 1971). St Patrick's Church in Grosvenor Street (1840-44) is now the oldest Catholic Church in the city, while the fate of the enormous Pitt Street Congregational Church (1841-57) is still undecided. With the completion of the Great Synagogue in 1878 all major religious groups had at least one substantial place of worship in the city centre. By the 1880s few city residents or visitors would have been more than ten minutes walk from an Anglican or Catholic Church.

The city's first school was built on what is now the junction of Castlereagh and Hunter Streets in 1793. It ended ignominiously when convicts set fire to it. What might be regarded as the first government school was opened at Parramatta in 1810; and in 1836 a Church of England School for the sons of gentlemen, the King's School, was opened at both Sydney and Parramatta. Students attending that school can still be seen wearing their 19th-century paramilitary regalia.

Private schools tended to follow their clientele to more congenial environments, and King's was typical in eventually abandoning its City site. The major exceptions are Sydney Grammar, founded in 1857, at what is now an educational and museological sanctuary on the edge of Hyde Park, and SCEGGS Darlinghurst, situated uncomfortably near to major thoroughfares and working-class terraces. Some small Catholic convents were established in the city centre and St Mary's Cathedral had a boys' school attached (as did St Andrews), but the prestigious Catholic schools, like their Protestant counterparts, preferred suburban sites with playing fields.

Government schools were built in large numbers after the Public Instruction Act of 1880 required the colonial government to provide for the education of all children. Fine sandstone schools can still be seen in many of the inner suburbs. Some public schools had already been built in the city. Fort Street School, established in the military hospital in 1818 and remodelled in 1849, became in the early 1850s the colony's leading elementary school and the main centre for teacher training. Secondary education was introduced in 1854. In 1882 Sydney Boys' and Girls' High Schools were opened in a building on Elizabeth Street which had formerly served as a church school. It was demolished in 1921 (to make way for David Jones), by which time single sex campuses had been established at separate locations in Moore Park (Burnswoods and Fletcher 1980).

Sydney began to acquire the adornments of a cultural capital in the 1860s. By the end of that decade the spires of churches and cathedrals and the clocktower of the Town Hall dominated the central spine of the city, flanked on its outskirts by substantial warehouses or parklands with wharves beyond.

The churches, built to last and protected by tradition, have proven more resilient to change than the schools, which have usually been demolished or turned over to other purposes. (The Fort Street buildings at Observatory Hill, for example, now house the National Trust.) Many suburban churches, however, have succumbed to declining congregations and been given new, usually commercial rather than residential uses.

The Australian Museum, founded as the Colonial Museum in 1827, moved to its present site in the 1860s. The earliest buildings at the University of Sydney date from the 1850s, while the present Art Gallery was begun in the 1890s. A Sydney Technological Museum was founded in the Garden Palace in the Sydney Domain. After the palace was destroyed by fire (in 1882), the Museum set up again in Ultimo in 1893 under the name Technological, Industrial and Sanitary Museum.

By 1900 Sydney had all the attributes of a major world city, for both residents and visitors, whether they came from the suburbs, the rest of the state, interstate or overseas. Public education to the age of 12 or 14 was freely available to all the city's children, although those living in overcrowded and unsanitary conditions were often too ill to avail themselves of it. Churches had been built by all the major religions although the Chinese were not well served.

In the 20th century the city has remained the centre for major museums and galleries and it still houses the most prestigious places of worship for all but the newly introduced religions (such as Greek Orthodox and Islam). But in the matter of education the City is no longer pre-eminent, although it continues to house the oldest University and, at the southern edge of the Central Business District, the new University of Technology (Sydney), first established as an Institute of Technology in 1965. Two other universities and more than a dozen colleges of advanced education have been built in the suburbs since the late 1940s, and western Sydney is finally to get a university of its own. The city area, reflecting its dramatic loss of population since the 1920s, no longer houses any government primary or secondary schools. Only a handful of religious schools and the non-denominational Sydney Grammar School hang on tenaciously, under threat of being sold up for their valuable real estate.

An Immigrant Population

Before World War II most Australians regarded themselves as of British stock, although some descendants of Irish settlers were not keen on the tag "British". From the late 1940s on, successive waves of immigrants have turned Australia—and Sydney and Melbourne in particular—into a remarkably multicultural society. In 1947 only thirteen per cent of Sydney's population was born overseas. But by 1990 almost one million people, or more than one quarter of its population, were overseas born.

In the 1950s and 1960s most of the migrants came from Europe, with English, Irish, Italians, and Greeks representing the largest groups. Italians, Greeks, and other Europeans made their presence felt in the inner suburbs, where they refurbished run-down terrace houses and opened shops and cafes. Many of these immigrants worked for low wages in factories. Some could command higher wages by working on projects in isolated areas, like the Snowy Mountains Hydro-Electricity scheme south-west of Canberra.

In the 1970s and 1980s the main sources of migrants were New Zealand, Yugoslavia, The Lebanon, and Asia, particularly Vietnam. Thus, by 1990, Sydney had as great a variety of ethnic retailing and restaurants as any city in the world. The traditional Chinese cafes of the interwar years, mainly run by descendants of mid-nineteenth century goldrush immigrants, now compete with Indian, Lebanese, Vietnamese, French, Italian and Greek restaurants. But apart from changing the Australian palate, these immigrants and their children have also changed the way Australians think about themselves. Sydney is no longer a white Anglo-Saxon outpost of the British empire, but a multi-racial society which has so far remained relatively free from racial violence.

Sydneysiders

Large cities tend to swallow individual personalities: the lives that are remembered in the history of Sydney are mostly those of the powerful or corrupt. Both good and venal politicians have made their way in Sydney in almost equal numbers. Criminals—who have controlled much of the city's gambling, drug trade and prostitution—have on occasion been aided by corrupt police. The distrust of authority figures that the average Australian is alleged to have, is reflected in widely-voiced doubts about the NSW police force. Over recent years, though, it has attempted to purge itself of dishonest elements.

The continuing push for development by both Labor (1976-1988) and Liberal (from 1988 onwards) Governments have created some bitter battles over the last two decades to preserve what is special about Sydney. The Harbour, foreshore bushland, historic buildings, and—unnecessary—freeways have been at the centre of most of the disputes. They also created some unlikely allies. In the 1970s, Jack Mundey, the working-class leader of the Builders' Labourers Federation, found himself on the same platform as author Patrick White, both protesting against the threatened encroachment on Centennial Park by development in the eastern suburbs.

Beyond the day-to-day workings of the city, Sydney's best-known citizens are the writers and artists who have attempted to capture its moods in prose, poetry, drama, photography, or painting. Christina Stead's *Seven Poor Men of Sydney* and *For Love Alone* include portrayals of the city in the inter-war years, whilst Patrick White's *The Tree of Man* reflects on life on the urban fringe, where the suburbs meld into the country. The paintings of Arthur Streeton, Lloyd Rees, Grace Cossington Smith, Margaret Preston, and Brett Whitely have presented many aspects of the form and feel of the city over the past one hundred years.

PART III
Sydney City Centre

1   Circular Quay

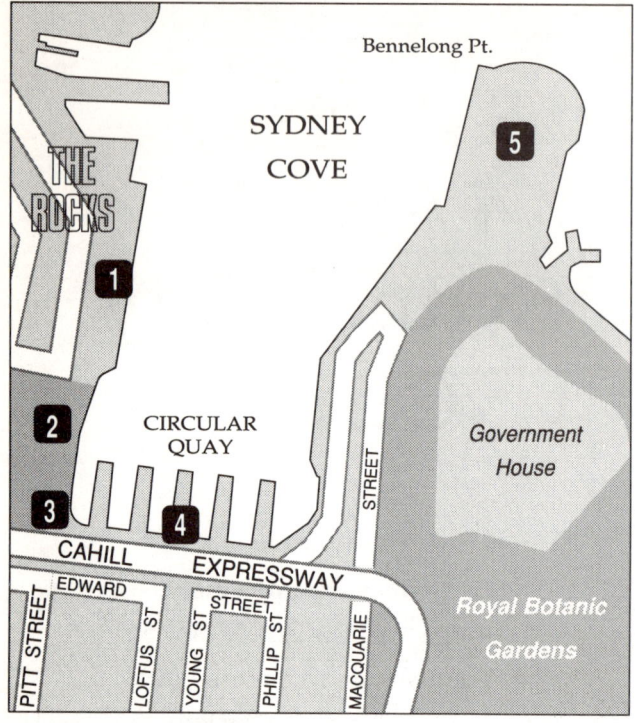

SYDNEY
COVE

Bennelong Pt.

THE
ROCKS

CIRCULAR
QUAY

Government
House

CAHILL

EXPRESSWAY

Royal Botanic

Gardens

1 Overseas Passenger Terminal
2 Museum of Contemporary Art
3 Tourist Information
4 Circular Quay Railway Station
5 Sydney Opera House

0   100   200 m

Circular Quay is the transport hub of the city—Sydney's front
door to both the city and the harbour. Catch ferries here to all
destinations. The railway station is Circular Quay; trains go
direct to Central Station, Wynyard and Town Hall Stations.
Many public bus routes begin at the Quay and a taxi rank is
just out front on Albert Street near Phillip Street. Most of the
boat and bus tours also start at the Quay. The airport bus leaves
from here, at the east end of the Quay close to Jetty 1. The Quay
is a few minutes' walk from many of Sydney's sights—a 10-
minute walk to the **Opera House**, a 5-minute walk to the
Botanic Gardens, and less than 5 minutes to the **Rocks**.
Sydney's main shopping area is a 15-minute walk down Pitt
Street.

The Quay is quintessential Sydney: Sydney commuters
heading to work mingle with visitors who have come down to
the Quay to enjoy the atmosphere. The Quay is also the centre
for street theatre. Fire eaters, jugglers, monocycle riders, pavement artists, and musicians, combine with the hoots of
departing ferries to bring the Quay alive.

Also known as Sydney Cove, this is the birthplace of
European settlement in Australia. Captain Phillip sailed into
the Cove on January 26, 1788 with 11 ships and 1300 people,

Replica of HMS Bounty at Circular Quay (*Ross Brownscombe*)

most of them convicts. By the 1850s Sydney Cove would have
been a scene of magnificent wool clippers with the quayside
lined with woolstores and warehouses. Now the clippers and
woolstores are gone and little remains to remind of those days.
Much of the Quay is a product of the 1950s and 60s, but thanks
to a badly needed facelift in 1988 the Quay is much cheerier
and open to its nautical surrounds.

If you want to explore the harbour, check Part 5 for some
ideas. The Quay is a good place to stop for an ice-cream or
coffee, cakes, donuts, hamburgers, chips, and doner kebabs.
Most of the cafes have sit-down service on the jetties, or you
can take-away. Opposite the jetties are restaurants with outside tables. You might like to sit out in the sun at "Rossini's"
Italian restaurant, with a drink or a plate of pasta. Or breakfast
at "City Extra" with your morning paper. Around the corner, get
fish and chips to take away for a picnic in the nearby Botanic
Gardens or hop on a ferry to eat them.

The shops at the Quay cater for both visitors as well as
commuters heading home. The souvenir shops are squeezed in
amongst the newspaper stands, fruit barrows, bread shop,
chemist and bottleshop. The **Rail Travel Information Centre**
will help out with rail trips out of Sydney. You can also book
bus tours, accommodation, and car rentals. Opening hours are
9 a.m. to 5 p.m. Monday to Friday, and 9 a.m. to 3 p.m.
Saturdays. There is a **currency exchange** here, open 8 a.m. to
8 p.m. Monday to Saturday, and 9 a.m. to 6 p.m. Sunday. There
is also a **tourist information booth** just to the west of the ferry

Johnny Marshall playing the didgeridoo at Circular Quay
(*Ross Brownscombe*)

terminal, towards the Rocks. Opening times are 8.30 a.m. to 5
p.m. Monday to Friday, and 8.30 a.m. to 4.30 p.m. at weekends.

Back from the Quay are the city's big buildings. The **AMP
Building**, directly behind the Quay, replaced one of the last of
the 19th-century woolstores. When completed in 1962, at 26
storeys high, it was the tallest building in Sydney and the
ultimate in modern design. From 1912 to 1957, the Height of
Buildings Committee kept buildings in Sydney to a maximum
of 48 metres. The AMP building was the first to take advantage
of the abolition of the height limit. Another product of the 1960s
was the Cahill expressway, the first stretch of inner city
freeway, completed in 1962. Many tourists, as well as Sydneysiders, wonder why an overhead freeway cuts across
Sydney's most spectacular view. Although controversial at the
time, citizens' groups had yet to discover community power! So
now the city is stuck with the cheap-and-nasty option.

Customs House, on the west side of the AMP, is a six storey
sandstone building built in 1888. The clock was added in 1897,
complete with dolphin and trident.

If you walk west around the Quay, towards the Rocks and
away from the Opera House (keeping the harbour on your right),
you will come first to a bulky sandstone building. This is the
old Maritime Services Board Building constructed between
1947 and 1952. In 1988 the Maritime Services Board sur-
rendered their building to Sydney's newest art gallery the
Museum of Contemporary Art.

Keep walking west along the Quay and you will come to the
Overseas Passenger Terminal, opened in 1961 to cater for
visitors to Australia arriving by boat. The large terminal was
soon rendered next to obsolete with the advent of jumbo jets.
In 1987 the scale of the terminal was trimmed down to cater
for the occasional cruise ship. As part of the recent facelift of
the Quay, the terminal was opened up and given a post-modernist treatment. It's a fine and unhurried place to watch the
ferries leave the Quay. A few restaurants have taken advantage

of the views: at the nearest end of the terminal is "Billy Blue's Kitchen"—sample Sydney's wonderful seafood in an informal atmosphere. At the other end is one of the famous Doyles seafood restaurants, "Doyles at the Quay". A number of tour companies are located here if you want to make a booking.

Walking east along the Quay, towards the Opera House (with the harbour on your left), the wall of buildings on the east side of the Quay is the product of a building boom in the mid-1950s. A row of old bond stores was replaced by the then ultra-modern 1950s office buildings. Developers are eying the opportunities for larger buildings, and the site is to be redeveloped. Already a couple of office buildings have been pulled down. A covered walkway leads to the Opera House. Half way along to the Opera House is the "Sydney Cove Oyster Bar". What finer spot to taste the wonderful Sydney Rock Oysters, and match this with a glass of Coopers draught beer. For a lighter snack, the "Portobello Cafe" has ice-creams and drinks, and quayside tables. Further along more serious eating is available at the lower level quayside walkway.

2   Sydney Opera House

In the early days of colonial settlement, cows grazed on this site. Between 1817 and 1819, a fort was built here to defend Sydney. Then, in the 1900s it became the site for a tram terminal. It's hard now to imagine the noise and grime of a tram terminal on this spectacular site. Yet spectacular as the site is, it is also demanding. Architects normally have to grapple only with the problem of creating one facade. Because the Opera House is visible from all sides and above, the architect, Jorn Utzon, had to puzzle over how to present five facades. What he has created is a building within a sculpture.

The distinctive shapes are evocative of sails, shells or even nun caps. Utzon explained his idea:

> If you think of a Gothic Church, you are closer to what I have been aiming at. Looking at a Gothic Church, you never get tired, you will never be finished with it—when you pass around it or see it against the sky. It is as if something new goes on all the time and it is so important—this interplay is so important that together with the sun, the light and the clouds, it makes a living thing.

In 1957, following an international competition, the then Premier of New South Wales, J.J. Cahill, proclaimed the Danish architect, Jorn Utzon, the winner. Initially, completion was scheduled for 1963 at an estimated cost of $ 7 million. Turmoil followed on a scale as grand as the concept. Costs blew out to $102 million and it was finally finished in 1973. The massive overrun was funded through State-operated Opera House lotteries. A good many of the problems stemmed from Utzon having submitted his inspired sketches to the competition without having thought through the engineering complications. Turning the sculpture into a working building needed a mammoth effort and demanded new technology. After endless quarrels with the project engineers, Jorn Utzon finally rolled up his sketches and walked off the job in 1966. Three Sydney architects took over the job—Peter Hall, Lionel Todd and David Littlemore. The interior design is mostly theirs.

The Opera House, now 17 years old, is already in need of refurbishment. The harbour has been eating away at the foundation piers; the roof needs repairing as the sealants have reached the end of their lifespan; some windows leak; and the interior requires upgrading as well. The delicacy and complexity of its construction demands constant maintenance.

Tours of the Opera House

Inside is a complete performing arts centre, with a concert hall, opera theatre, drama theatre, and playhouse cinema. Four restaurants, and eleven bars cater for casual visitors and patrons, together with four gift shops, and a library for students of the arts.

Guided tours of the Opera House suit all tastes and schedules. Choose from the following:

Guided tours of theatres, halls and foyers last 1 hour. Tour times are everyday, continuously from 9 a.m. to 4 p.m. The cost is $5.50, half price for all students and children, and Australian pensioners. For bookings phone 250 7250.

Backstage tours last 90 minutes. Tours are on Sundays only, continuously from 9 a.m. to 4 p.m. The cost is $8.50; phone bookings on 250 7250.

Mini-tour and lunch at the Harbour Restaurant, with a 20 minute tour followed by lunch. Tour times are 11.30, 12.30, 1.30, Monday to Saturday. The cost is $24.50. Ring 250 7447, or 250 7197 for bookings.

An **Evening at the Sydney Opera House** is a package providing dinner and a performance. Phone 250 7197 or 250 7198 for bookings.

Restaurants

The **Bennelong Restaurant** is the most expensive of the four restaurants. Monday to Saturday the Bennelong is open for lunch and pre-performance dinner. For bookings, phone 250 7578. Seafood is the specialty of The **Harbour Restaurant**. You can eat outside along the harbour wall, phone 250 7191. It also has take-away fish and chips. The **Forecourt Restaurant** is a brasserie-type restaurant and has cocktails and coffee. **Cafe Mozart** is a theatre cafe and open before and after performances.

The Performing Arts at the Opera House

This list gives an idea of the approximate seasons for the performances. The box office number is 250 7777. Reduced ticket prices are available for Australian pensioners and children. Half price tickets for students are available for selected performances. For example, the Australian Ballet offers students half price on Monday nights and the 6.30 p.m. performances. Look in the *Sydney Morning Herald* for details.

Ballet: March, April, May, and December

Opera: all through the year, except April, November, and December

Sydney Symphony Orchestra: March to November.

Sydney Philharmonia: March to November

Sydney Dance Company: April to May

Light Opera: February and March

---

3   Bridge and Macquarie Streets

These two streets are behind the Quay and are two of the most historic of the city's streets.

Bridge Street

Just one block up from the Quay is Bridge Street. Always an important street, it formed the administrative centre of the early colony.

The little handkerchief park on Bridge Street known as **Macquarie Place** is packed with history. Once part of the gardens of First Government House, it later became a public square surrounded by orchards and stately buildings. The streets and buildings squeezed the square down to its present tiny size. The obelisk (1818), marks the point from where all the road distances from Sydney to other settlements in the colony were measured. The names of the country towns carved in the obelisk show the extent of the then tiny colony. The bow anchor and cannon come from the **HMS Sirius** which escorted the First Fleet. The cannon was landed shortly after the founding of the colony and was used to signal the arrival of ships. The bronze statue is of Thomas Mort, who died in 1878. Mort was instrumental in establishing the wool trade in Australia, and pioneered refrigeration for food export. Two trees were planted here by Queen Elizabeth II and the Duke of Edinburgh in 1954, marking the first visit to Australia by a reigning monarch. The first street lamp in Australia stood here in 1826 and burned whale oil. It has been replaced by three gas lamps. Sydney could sorely do with a few more of these lovely little parks to offer relief from the concrete and heat of Sydney's streets.

Fronting on Macquarie Place, tucked behind some palm trees is **Kyle House**. Designed by one of Sydney's leading art deco architects, Bruce Dellit, it was constructed in 1931. Two special features are a lovely brickwork frieze of ceramic tile at the top, and the grandiose staircase inside the arched entry. Dellit's most famous building in Sydney is the War Memorial at the southern end of Hyde Park.

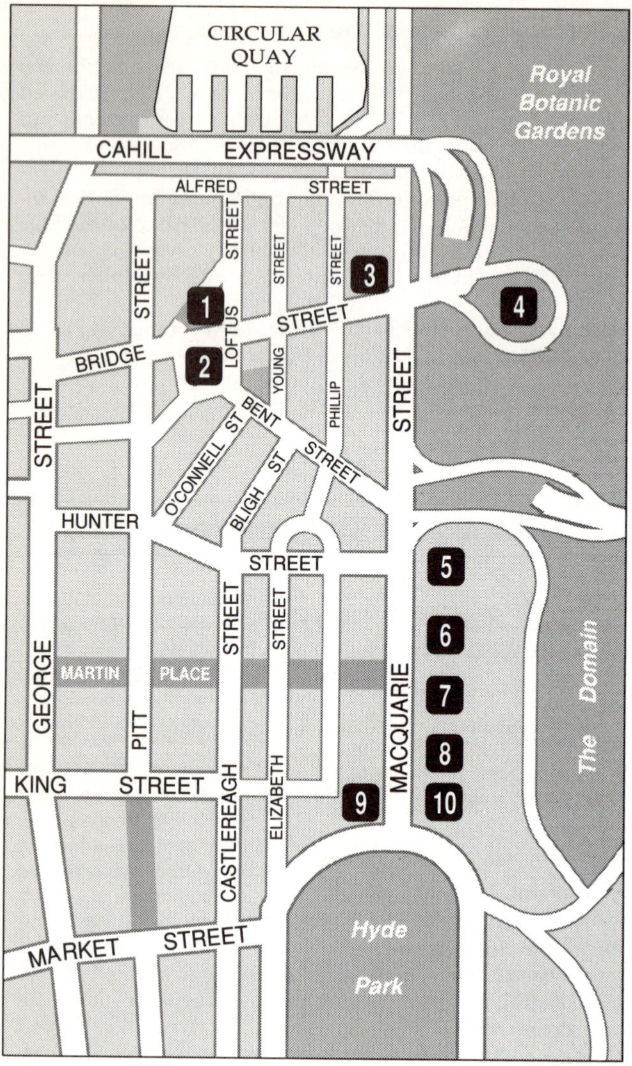

CIRCULAR QUAY

Royal Botanic Gardens

CAHILL EXPRESSWAY

ALFRED STREET

STREET

STREET

STREET

STREET

1

LOFTUS

3

STREET

4

YOUNG

BRIDGE

2

PHILLIP

STREET

BENT

O'CONNELL ST

STREET

STREET

BLIGH

ST

STREET

HUNTER

MACQUARIE

STREET

5

STREET

STREET

6

GEORGE

MARTIN   PLACE

7

The Domain

PITT

8

KING   STREET

9   **10**

CASTLEREAGH

ELIZABETH

Hyde

MARKET   STREET

Park

Bridge and Macquarie Streets

1 Macquarie Place	7 Sydney Hospital
2 Lands Department Building	8 The Mint
3 Intercontinental Hotel	9 St. James Church
4 Conservatorium of Music	10 Hyde Park Barracks
5 State Library of N.S.W.	
6 Parliament House (N.S.W.)	

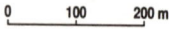

0   100   200 m

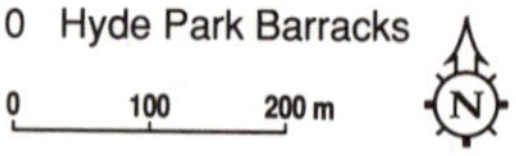

On the other side of Macquarie Place is the impressive **Lands Department Building**. This three storey classical revival building (1877-90), was designed by the Government Architect, James Barnet. The exterior of the building has 48 niches for explorers and politicians "responsible for promoting settlement". Twenty-five of these niches are still vacant! Note the beautiful onion-shaped clock tower. Inside the front foyer is a Department of Lands **map shop** where maps, posters, aerial photographs, and books are for sale. Open weekdays 9 a.m. to 4 p.m.

The **site of the First Government House** is on the corner of Bridge and Phillip Streets. The House was demolished in 1845 but its foundations were recently covered over to help preserve them. The foundations were laid in 1788 and was originally a modest six-room house. However, it was enlarged over the years and housed nine colonial governors. A new Government House was built in the Domain in 1845. The **Chief Secretary's Building** on the corner of Bridge and Phillip Streets, was also designed by James Barnet, around 1878. The style is strongly reminiscent of many Parisian buildings of this period. The former **Treasury Building**, at Bridge Street and Macquarie Streets, was designed by Mortimer Lewis in 1849 in the Renaissance style and later extended in 1894. In 1986 the Treasury was incorporated into the "Intercontinental Hotel", one of Sydney's most expensive hotels. A look around the lobby reveals the former Treasury Courtyard. The Cortille Cafe is a favourite stop for coffee and cakes and drinks, at five-star hotel prices of course.

Macquarie Street

Governor Macquarie was responsible for laying out this lovely avenue and most of the splendid 19th century buildings found along it. Thanks to Captain Phillip it is framed on one side by a continuous expanse of green space, from the Royal Botanic Gardens to Hyde Park. Macquarie Street is like a breath of fresh air after the dull and dingy canyons of much of the business district. Macquarie proclaimed the street in 1810 and its north-

The Macquarie Street entrance to the Botanic Gardens

ern end reached only as far as Bent Street. Not until 1845 was it extended to the harbour. In 1914 the road pavement was widened by 6 metres, and the majestic fig trees along its length were removed. The Director of the Botanic Gardens decided to line the street with Canary Island Palms to convey "a more semi-tropical aspect".

The **Conservatorium of Music** is the large building set back in the gardens, facing Bridge Street. Designed by Francis Greenway and completed in 1821, the original building was designed as stables for the new Government House Governor Macquarie planned, and was modelled after Inverary Castle in Scotland. Imagine how large the house would have been! Plans for Government House were scaled down to more modest proportions. As reliance on horses declined, the building was converted in 1913 to the Conservatorium of Music. The Conservatorium is open daily from 9 a.m. to 4 p.m. Free concerts are held on some Wednesdays at 1.10 p.m. during school terms. Friday evening twilight concerts are also held at 6 p.m., for which a modest admission is charged. Ring 230 1263 for bookings and information.

The lovely Victorian townhouse at No. 145, gives us an idea of Macquarie Street in its elegant prime. Originally only two storeys, the upper storeys were added in the 1900s. Another gem is the art deco **British Medical Association House**, 135-137 Macquarie Street. Designed by Fowell and McConnell in 1928, it is decorated with a playful mix of medical and Australiana motifs in glazed terracotta blocks. Across the street, is the Palace Gates entrance to the Royal Botanic Gardens. The kiosk at the gates serves salads, sandwiches and drinks and is a pleasant place to stop on Macquarie Street to admire the view.

On the corner of Macquarie and Bent Streets, the State government built in 1967 this 38 storey building, making it the tallest office building in Sydney at the time. The office of the Premier of New South Wales is in this building. The old library, an elegant stone building dating from 1845, was demolished to make way for this tower.

The **State Library of New South Wales** occupies the imposing sandstone building facing the Botanic Gardens. It was
originally built in 1910 but several additions have been made
including the fabulous new wing facing Macquarie Street,
completed in 1988. The old wing houses the enormous collection of Australiana, called the **Mitchell Library**. The new wing
houses the State Library's main General Reference library and
has three million items. The facilities must be the envy of
libraries around the world. For example, one special area caters
for people tracing family histories with a huge collection of
records of births and deaths, shipping arrivals, etc. In another
area a high-tech reading machine is used to convert printed
material into speech and Braille. The library's special historical
exhibitions are a popular attraction. There is a restaurant and
gift shop. Reference library hours are 9 a.m. to 9 p.m., Monday
to Friday, 9 a.m. to 5 p.m. Saturdays, and 2 p.m. to 6 p.m. on
Sunday. Phone 230 1414 for more information.

Parliament House and the **Mint Building** were two of the
original wings of what was known as the "Rum Hospital" built
in 1816. In a scheme with a familiar and topical theme,
Governor Macquarie granted enterprising businessmen a
monopoly in rum imports for three years, and the right to import
45,000 gallons, in return for the construction of the hospital at
"no cost to the colony". Short on funds, the proposition ap-
pealed to Macquarie. The only cost to the government was the
supply of some convict labour, oxen, and bullocks. Although it
still stands today, critics claimed upon its completion that due
to the shoddy workmanship and inferior materials the building
would soon fall to pieces, and Macquarie resolved never again
to enter into shonky deals with developers.

In 1816 the Waterloo Ball was held in the Rum Hospital to
celebrate the British victory (it took seven months for the news
of the victory to reach the colony). Parliament moved into this
building in 1829 and the building has been altered several
times over the years. Open times are 9 a.m. to 4.30 p.m.,
Monday to Friday. Admission is free. To see the Parliament in
session, go to the front desk on Tuesday and Wednesday at 2
p.m. or Thursday at 10.30 p.m. Parliament generally sits from
mid-February to mid-May and from mid-September to early
December. Phone 230 2111 for more information.

The **Sydney Hospital** is the large Victorian structure
dominating the east side of Macquarie Street. In the 1960s the
Government Architect's Office recommended demolishing the
hospital and moving the beds to Western Sydney where they
were badly needed. Parliament was to have a new building on
the hospital site. However, the medical profession was successful in stopping the move. In the early 1980s the Minister for
Health in the Labor Government took on the doctors again and

drastically reduced the functions of Sydney Hospital, but retained the building. The fountain of the wild boar in front of the hospital is a replica of *Il Porcellino* fountain in Florence.

After the Gold Rush, a branch of the **Royal Mint** was established in the handsome south wing of the Rum Hospital. Established in 1853 and closed in 1927, the name "Mint Building" lives on. In the 1980s the Mint building was restored ' and reopened as a museum for decorative arts, coins, and stamps. Admission is free. Opening hours are listed in Part 8.

The **Hyde Park Barracks** at Queen's Square, Macquarie Street (1819), was designed by the convict architect, Francis Greenway. Governor Macquarie was so impressed with the elegant design, he granted Greenway a full pardon on his forgery conviction. The convict barrack housed up to 1,000 convicts at one time. Supposedly street crime was reduced to one-tenth its former proportions once convicts were safely tucked away at night. Prior to the completion of the barracks, convicts were free to stay wherever they could find accommodation. The clock is the oldest in Australia. The Barracks is now a museum of social history and the upper floors show the original convict quarters complete with their hammocks. During renovations around 1980, a complete convict shirt was recovered from under the floorboards. Reproductions of the striped shirt are available in the Barracks gift shop. Admission is free. Opening hours are listed in Part 8. The pleasant restaurant in the courtyard of the Barracks, is popular at lunch times so go early on weekdays.

St James Church on Queen's Square, Macquarie Street (1820) is another of Greenway's graceful buildings. Originally intended as law courts, the design was changed to a church. The simple lines of the exterior are matched by the interior. Plaques commemorate early citizens, including many explorers. Having turned the court rooms into a church, The **Supreme Court Building** across from the church was redesigned from a school to law courts and completed in 1828. In the early 1970s this building was under serious threat of demolition but was fortunately spared. The big, high rise building beside St James Church is the **Law Courts Building**. On the 14th floor is a reasonably priced cafeteria opened to the public (weekdays only). But the real reason for coming here is the glorious views over Hyde Park, the city, and Woolloomooloo. Breakfast is served from 7 a.m. to 9 p.m., lunch, 11.30 to 2 p.m., and snacks and drinks until 3.30 p.m.

4   The Rocks and Millers Point

*To reach the Rocks, take the train to Circular Quay Station, or
the 431 bus along George Street.*

The Rocks is a delightful place to explore, with a definite
sense of history. It is the oldest area of European settlement in
Australia, the first buildings erected a few weeks after The First
Fleet arrived in 1788. Around every corner there's another
surprise—a breathtaking view of the Opera House or Harbour
Bridge, a quaint old building down a narrow alley, or a 19th
century pub. The Rocks is a great place for a meal, a drink, or
to shop for Australiana souvenirs.

The Rocks has showed its remarkable resilience by surviving
many invasions over the years. From its earliest days the Rocks
was taken over by sailors. For their entertainment, dozens of
taverns sprang up, together with brothels, gambling houses,
and opium dens. By 1810 there were 50 beer licenses issued
in Sydney, most of them in the Rocks. Some early pubs had
colourful names, such as "the Sheer Hulk" and "Black Dog".
The Lord Nelson, the Hero of Waterloo, and the Orient remain
from the 1800s and are a special treat if you enjoy pubs with
character.

The next invaders were rats spreading "the Black Death",
bubonic plague. Australia's first major slum clearance project
quickly followed in 1900. Large areas of the Rocks were
demolished, roads widened, and houses whitewashed to stamp
out the disease. The Harbour Bridge cut a great swathe through
the Rocks in the 1920s, as did the Cahill Expressway in the
1950s. Even in a rundown condition, developers recognised a

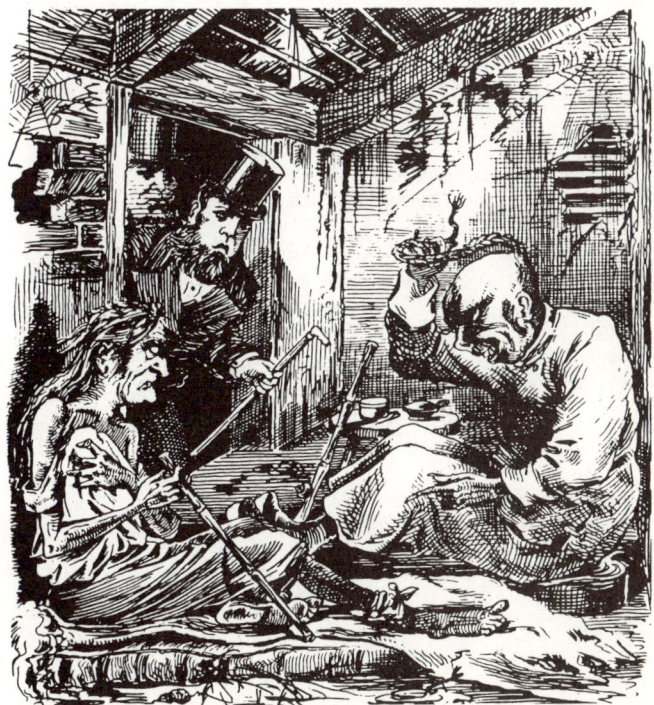

Cartoon of opium den, the Rocks (*Sydney Punch*)

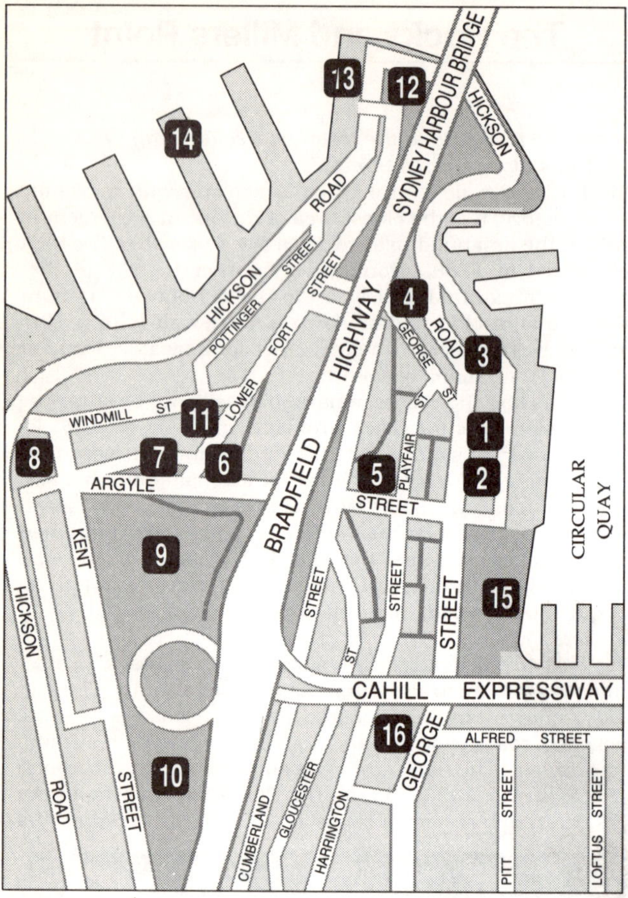

The Rocks

1   Rocks Visitors Centre
2   Cadman's Cottage (National
      Parks Information Office)
3   Campbell's Warehouse
4   The Earth Exchange
5   Argyle Centre - Argyle Steps
6   Garrison Church
7   Argyle Place
8   Lord Nelson Hotel
9   Sydney Observatory

10   National Trust Centre
11   Hero of Waterloo Pub
12   Dawes Point
13   Pier 1
14   Wharf Theatre / Restaurant
15   Museum of Contemporary Art
16   Regent Hotel

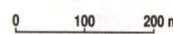

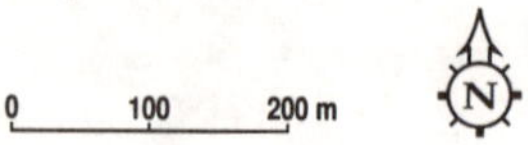

prime site when they saw one. In the race to build, it became a tempting opportunity. Yet, against all the odds, proposals to flatten the Rocks and replace it with shiny, new high-rise office buildings and hotels were stopped just in time. By the early 1970s the remaining residents organised for community action. With the help of the builders labourers' union, which issued work bans on heritage sites and workers' homes, the Rocks was saved from wholesale redevelopment.

The Rocks is now controlled by the **Sydney Cove Redevelopment Authority**. The changes have been dramatic: a mixture of restoration, conservation, and both sensitive and insensitive redevelopment has transformed the Rocks. Some people may object to this—and it's true, the area now does have an overly polished look in parts, with an overlay of trendy shops and restaurants. Still, redevelopment and investment in new businesses have revitalised the area. And what saves the Rocks, the essence of its special character, is the preservation not just of old buildings, but of something of the old way of life. The local community retains a precarious hold. And whilst it survives, the Rocks will overcome this latest invasion, this flood of souvenir shops and stuffed koalas, and continue to fascinate.

There are two distinct areas, separated by the Bridge: on the western side are Millers and Dawes Points; and on the eastern side, opposite the Opera House, is the Rocks.

The Rocks, George Street

George Street in the Rocks is the oldest street in Australia and even now has a charming 19th-century streetscape, brought to life by a wonderful mix of shops, pubs, and restaurants. You might like to start your visit to the Rocks by calling into the **Rocks Visitors' Centre**, in the old coroner's court at 104 George Street. Don't miss the short free film describing the history of the Rocks. You can pick up some extra information on the Rock's attractions as well as a good free map. Cassettes are available for hire which give you an fascinating self-paced tour of the Rocks. Opening hours are 8.30 a.m. to 5.30 a.m. Monday to Friday, and 9 a.m. to 5 p.m. Saturday and Sunday.

Next door to the Visitors Centre is the **Marionette Theatre**, 106-108 George Street (1865). This sandstone building of Victorian Romanesque style was originally a hostel for visiting sailors. It now houses the Marionette Theatre of Australia. The **Craft Centre**, at 100 George Street (built in 1909), occupies a building dating back to 1857 which was originally the seamen's mission. The neo-classical facade was added in 1909. Inside are many exquisite items, including modern pottery and ceramics, glass, wall hangings, materials, clothes, cards and jewellery. This is a perfect place for souvenirs.

Further along at 66-84 George Street are the **Metcalfe Bond Stores** (1912-1916), now used for galleries, shops and restaurants. Australian designed goods on display include hand-tufted carpets, furniture and sheepskin products. There is also a decorative glass maker. Before George Street disappears under the Bridge is the **Earth Exchange** at 36 George Street. This building was built as a DC power station but never used for that purpose, as the change to AC came before it became operational. Formerly called the Geological and **Mining Museum**, it has been completely renovated and enlarged to house an exciting new collection on energy, mining, and geology. See Part 8 for opening hours.

Restaurants lining the waterfront at the Rocks

On the other side of the street at 25 George Street is the **Mercantile Hotel** (1915). This pub has lovely art nouveau details, and the tiling on the outside is traditional of pubs of this era, now sadly disappearing from Sydney. Next door is the **Sea Shell Shop**. This tiny treasure trove probably has more shells than you have ever seen together in one place before: shells in all shapes and sizes, hard and soft corals, sharks' jaws and swordfish swords. This is a real treat, with every nook and cranny stuffed with fishy oddities.

At 39 George Street in a Victorian terrace is the **Nature Conservation Council Shop**. Amongst the stuffed koalas, tea towels and posters are some attractive carved banksia vases. Next along George Street is the old Bank of New South Wales, lovingly restored, and still operating as the Westpac Bank, open normal banking hours. Behind, in a new extension is the **Westpac Museum**, where you can experience a working branch of the 1890s. See Part 8 for opening hours. Admission is free. Right behind this is Atherton Place, reputed to be Sydney's shortest street. The terrace houses come to an abrupt end in a bleak and clearly unyielding cliff face.

The "Orient Hotel" at 89 George Street (circa 1850), is one of the oldest hotels in the Rocks and has been operating since 1850. This next stretch of George Street, as far as the Cahill expressway at the Quay, is the most charming, lively and original collection of shops and pubs and restaurants in the whole of the Rocks, perhaps the whole of Sydney. A few of the places to explore are delightful pubs like the "Fortune of War" (No. 137) and the "Observer" (No. 69) and shops such as the "Graphics Factory", "Weiss Art at the Rocks" and "Ken Done",

"Australian Wildflower Specialists" (No. 79), and the "Australian Heritage Bookshop" (No. 81 1/2). The Old Rocks police station (No. 127), complete with lock-ups, built in 1882 on the site of Australia's first hospital, has been transformed into craft displays, including ceramics, leather, wood and clothes. It's "Morrison's" at 105 George Street for Australian bush clothing such as Akubra hats and stockman's oilskin coats. Stop to enjoy the experience, perhaps with a coffee at the "G'Day Cafe" (No. 83), or if it's lunch time, consider barbecuing yourself a fat steak at "Phillip's Foote". There's a Japanese Deli on Nurses' Walk, just behind the Russell Hotel.

The **Russell Hotel** (No. 143) is a late Victorian building after the "tudor revival" style and has a lovely circular corner tower. Lastly, across the road is **Cadman's Cottage**, 110 George Street (1815-16). This simple stone cottage is Sydney's oldest dwelling and was the home of John Cadman, a convict sentenced to death for stealing a horse. Spared from death, he was transported to Sydney in 1798, pardoned in 1814 and became Government Coxswain in 1817. The cottage now houses the **National Parks and Wildlife Shop** where you will find interesting publications for sale, and a large collection of free maps and pamphlets of the national parks in the Sydney Region. Opening hours are 9 a.m. to 5 p.m. Mondays to Fridays, 9.30 a.m. to 5.30 p.m. on Saturdays and Sundays.

The Rocks, Hickson Road

Hickson Road joins George Street at Metcalfe Stores, and leads down the hill to Dawes Point, under the Bridge. Hickson Road continues around the Point through Millers Point on the far side of the Bridge.

The major building on Hickson Road is **Campbells Warehouse**, 11-31 Hickson Road and 7-27 Circular Quay West (1842-1861). Robert Campbell established the first commercial wharf in Australia and was one of the early giants of commerce. Now the beautifully restored warehouses offer a lovely setting for restaurants. What finer setting for a leisurely lunch? From

Cadman's Cottage

the outside tables you may enjoy breathtaking views of the harbour and Opera House. The restaurants include: the "Waterfront Restaurant" a seafood restaurant complete with old schooner masts, the "Imperial Peking" considered one of the finest chinese restaurants in the city, and at the far end of the Warehouse, the large Italian restaurant "The Italian Village". "Pancakes on the Rocks" is accessed from the Hickson Road side.

Worth a visit is the **Australian Wine Centre**, on the Quay side of the Warehouse. It has a wide selection of fine Australian wines. The Centre will pack and export any purchases on request. Whether or not you're in the market for rugs don't miss the "Halcyon-Lake Carpets", on the Hickson Road side. This small store has stunning rugs from a number of countries, including Australia.

The Rocks, Argyle and Harrington Streets

Argyle Street joins George Street at the Orient Hotel. Argyle Street, through the Cut, is the easiest access to Miller Point, and onto the Harbour Bridge freeway.

Heading up Argyle Street to the Cut, on the right hand side is the **Argyle Centre**, (Nos 18-20). The sandstone and brick warehouses were originally bond stores (original section 1828). Inside are four storeys of shops and galleries selling souvenirs of almost any description, including some fine selections of Aboriginal art and artefacts, and what surely must be the world's largest concentration of stuffed koalas under one roof.

The Solar Energy Information Centre on the third floor has displays on renewable energy, publications, books, as well as solar-powered gadgets for sale. The Total Environment Centre, provides information on the critical environmental issues of the day, and has a good selection of nature and conservation books on sale. If you continue to the very top of the centre you may be able to find your way out the back door to Gloucester Walk, and Cumberland Street, the upper tier of the Rocks, and access to the footpath on the Harbour Bridge.

Harrington Street continues on both sides of Argyle Street. In front of the Centre is Rocks Square with its much photographed **First Impressions** sculpture representing the first European settlers of Sydney. The quaint workers' cottages on the west side of Harrington are occupied by shops, take-aways and small restaurants. They open also onto a small passage behind, which adjoins the Argyle Centre.

The "Argyle Tavern", at 12-18 Argyle Street (1830), is a pub and restaurant offering Australian food and entertainment. See a sheep being shorn and enjoy gum leaf music at the cabaret. On the other side of Argyle Street is the new post-modernist style Clock Tower Square. This attractive development again is full of souvenir shops (including the National Trust Gift Shop), overflowing with stuffed koalas on every available shelf and nosing out from every cranny. Elsewhere on Harrington Street are the Standford Apartments, squeezed between small cottages, and the Harbour Rocks Hotel, an intriguing conservation and development project building in an old woolstore.

Before you leave this area, if you're feeling energetic head up Argyle Street to Argyle Cut. The **Argyle Cut**, a great road passage through sandstone was hewn to provide a road from Sydney Cove to Darling Harbour. It was begun with convict labour in chain gangs in 1843. After several years of slow

The Argyle Cut

progress it was abandoned and remained a pedestrian passage until 1859 when it was finally completed. The west side of the Cut is called Millers Point (described in the next section).

Immediately before the Cut, on the right hand side are the **Argyle Stairs** (1818). These steps lead up to Cumberland Street and from here, another set of stairs lead up to the Harbour Bridge walkway. Away from the tourist buses, this quiet area has a special charm. The short climb is well worth the effort. The view from the top looking down the stairs and across the old bond stores captures the real charm of the Rocks. Two

rather lonely pubs, "Gloucester Hotel" and "Australian Hotel", survive on Gloucester Street, and are a reminder of the Rocks before the gutting and reassembling of restoration projects. If you make it this far, walk onto the Bridge and view the harbour from here. Keep on going to the lookout and museum in the first **Pylon** (see Part 5 for details).

Back at the bottom of the Stairs, and opposite is Cambridge Street. This street was the hub of the Rocks squalor. The southern part of this road was obliterated by the Cahill Expressway—here the infamous "Sheer Hulk" pub once stood.

Millers Point, Argyle and Lower Kent Streets

Pass through the Argyle Cut into a totally different world. Few tourist buses make it this far, and stuffed koalas are completely unknown. Here lives the remnants of the old community, still clinging to prime development land. This is a very special little place with a wonderful village atmosphere.

The **Holy Trinity (The Garrison) Church**, Argyle and Fort Streets (1840-1878), was the first official military church in the colony. Plaques above the door and on church pillars show the regimental crests of the English garrison regiments that were on tours of duty in Sydney. Note the beautiful stained glass window above the altar. Edmund Barton, Australia's first Prime Minister attended the old school in the churchyard from 1846 to 1851.

Argyle Place Park, at Argyle and Lower Kent Street, is a wonderful patch of green surrounded by graceful terraces, and complete with gas lamps, a drinking fountain built in 1869, and a cast iron post box. The mid-19th century houses around the park are worth a close look. No. 50 is particularly charming, a Georgian cottage with lovely cast iron frieze and railings, shutters, and fanlight.

The **Lord Nelson Brewery Hotel**, 19 Kent Street, on the corner of Argyle Street (1832), has been licensed since 1842, and is the oldest hotel in Sydney. It brews its own boutique beers on the premises. Dan Quayle, Vice President of the USA, made an unscheduled stop here on his 1989 Australia visit. The beers carry tantalising names such as Nelson's Blood Stout and Fleet Wheat. Upstairs is the Lady Hamilton Restaurant. Bed and breakfast is also available.

The **Palisade Hotel**, 35 Bettington, at the end of Argyle Street is a pleasant pub and restaurant with accommodation above. Despite so many pubs in the area being spruced up, The Palisade remains in its original condition, and is popular with the port workers and the local community. "Fish at the Rocks", 29 Kent Street, is a comfortable BYO restaurant and serves delicious seafood. It's a favourite with office workers, so book in advance. Open Tuesday to Saturday for lunch and dinner (27 3137).

Millers Point, Observatory Hill

Sydney Observatory (1858), on Observatory Hill, can be reached from Kent Street by taking the steep Agar steps, near High Street. You can also get there by car up Watson Road, which is a ramp-like street leading off Argyle Street, west of Argyle Cut. Another access point is up the Argyle Stairs, to the Harbour Bridge stairs, and then through the subway to Obser-

vatory Hill. The view of the harbour is magnificent from here and the enormous Moreton Bay fig trees make a perfect, shady rest spot. The Observatory occupies the site of Fort Phillip (1804-05), and some remains of the fort can be seen. The Observatory has two domed observatories. The time ball on the top of the weather vane is a well-known Sydney landmark. It was devised to announce the correct time to ships in the harbour and the people of Sydney. The time ball is dropped precisely at 1 p.m. every day as a signal for a gun to be fired at Fort Denison for people to set their watches—much more romantic than radio beeps. This practice was initiated in 1858, was stopped during the war and only recently begun again. There is a permanent "hands-on" astronomy display in the Observatory. Night viewing of the stars tours are available every night. See Part 8 for more details. Bookings are essential (241 2478).

National Trust Centre on Observatory Hill occupies part of what was once a military hospital built under Governor Macquarie (1815). From the 1840s it was enlarged and became the Fort Street School. The National Trust is charged with looking after Australia's heritage. Opening hours for the Centre are 8.30 a.m. to 5 p.m. Monday to Friday, and 2 p.m. to 5 p.m. Saturday and Sunday. The S.H. Ervin Gallery here always has interesting exhibitions of Australia's environmental and artistic heritage. See Part 8 for gallery hours. The cafe serves Devonshire teas and lunch. There is also a bookshop with a fine collection of heritage books and souvenirs.

Millers Point, Lower Fort Street

Just off Argyle Place, this delightful street leads to Dawes Point at the northern tip of the Rocks. It also takes you down to the wharves in Walsh Bay. Lower Fort Street is one of the finest 19th century streetscapes. Unlike the shops on George Street this is a residential street with a mixture of Georgian townhouses, large Victorian terraces, as well as Regency-style townhouses. These stately homes contrast with the very modest workers' cottages found in the nearby Rocks. Many of the houses in this street have basements, which are a very rare feature of Australian houses. The famous **Hero of Waterloo** is at 81 Windmill Street, on the corner of Lower Fort Street (1844).

The cellar of this pub was used to lock up drunks, as witnessed by the bars in some of the lower windows.

The **Colonial House Museum** at 53 Lower Fort Street (circa 1880), gives an opportunity to look inside one of the stately Lower Fort Street homes. It's full of Victoriana. Open 10 a.m. to 5 p.m. Monday to Sunday. Another house of interest is **Bligh House**, 43 Lower Fort Street (1833), one of the finest house on the street. Note the Doric columns along the front made of wood. The house was built for Robert Campbell Jr. who was a son of Robert Campbell of wharf and bond stores fame. The Australian College of General Practitioners restored the house and have their offices here.

At the end of Lower Fort Street and under the shadows of the Sydney Harbour Bridge is another lovely harbourfront park, **Dawes Point**. The view to the Opera House and around to Circular Quay is stunning. You can walk all the way around to the Opera House from here along the harbour. Lieutenant William Dawes installed a gun battery on the point in 1788 to defend the colony.

Millers Point, Hickson Road

From Lower Fort Street, a steep set of stairs lead down into Hickson Road, and to the wharves on Walsh Bay. These nine wharves were built from 1910 to 1922 as part of the Government's clean up after the outbreak of bubonic plague in 1900. The wharves were used for wool bale storage, berthing of passenger liners, and general cargo handling. Most of the wharves were no longer in use by the early 1960s and were left undeveloped until the 1980s. Now like a set of dominoes, one after another is being redeveloped to take advantage of the wonderful location. In the next few years the whole area will be redeveloped for tourism, housing and leisure uses.

Pier 1 on No. 1 Wharf, Hickson Road, built in 1912 as an ocean passenger terminal, has been redeveloped along the style of Fisherman's Wharf in San Francisco. The wharf building has a collection of restaurants, fast food outlets, and shops. Somehow it just hasn't captured the nautical atmosphere and doesn't live up to its potential. The inside is rather congested but the outside offers pleasant wharfside tables. A free bus to and from Pier 1 travels to Circular Quay railway station.

Wharf Theatre on No. 4/5 Wharf, Hickson Road. This wharf building houses the Sydney Theatre Company specialising in plays by Australian playwrights. There is a cafe and fine restaurant located right at the end of the wharf, making it a great place to watch the boats go by.

5   Royal Botanic Gardens

The Gardens are a pleasant stroll from Circular Quay. There are several different ways of entering the gardens. A good place to start is the gates opposite the State Library off Macquarie Street.

Admission to the Gardens is free and the gates are open from 8 a.m. to sunset. Surrounding the Gardens on the east side and the south side is more open space known as the Domain. A lovely road with views over the Gardens and Woolloomooloo Bay cuts through the eastern Domain.

History of the Botanic Gardens

The Gardens area was set aside by Captain Phillip in 1788 to provide a farm for the new settlement. Only a small part of the area was actually farmed. Governor Macquarie began to establish the Botanic Gardens in 1816 and some of the planted trees are the best part of 200 years old. Mrs Macquarie's Road, leading to **Mrs Macquarie's Chair** on the eastern side of the Gardens was completed in 1816. The wife of the Governor is said to have instigated this road project and the ledge of rock where she used to sit keeps her name. Despite tourists by the bus load and the clicking of cameras, this lovely, tranquil perch can still be enjoyed by walking a few metres from the road. The most famous view of Sydney is from this spot, with the Opera House against the backdrop of the Harbour Bridge.

A Sydney resident in the 1820s gave this description of the western fringe of the gardens:

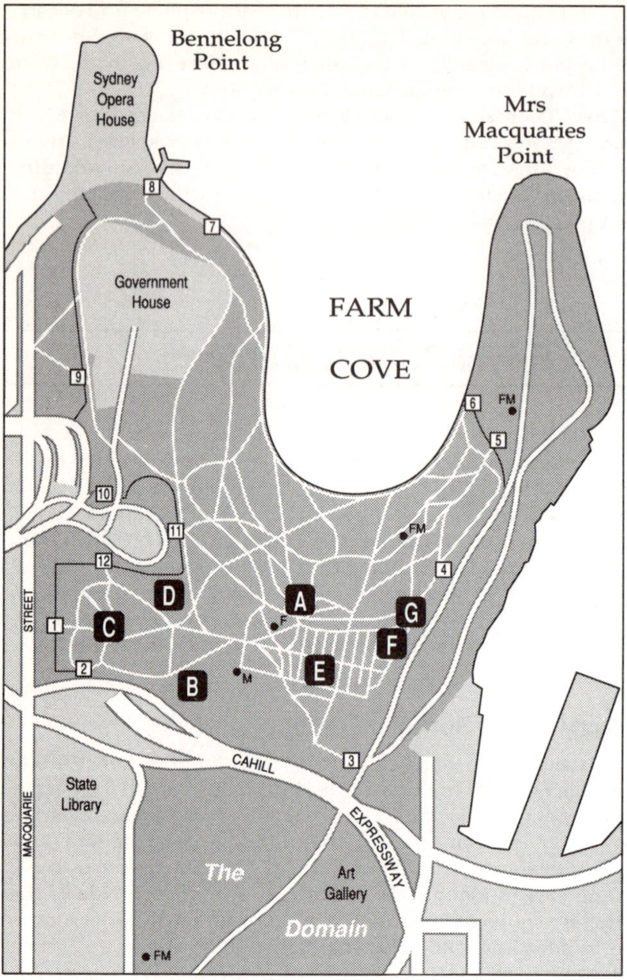

Bennelong
Point

Sydney
Opera
House

Mrs
Macquaries
Point

8

7

Government
House

FARM

COVE

9

6 FM

5

10

11

FM

12

4

D

A

G

STREET

1

C

F

2

M

E

B

3

CAHILL

State
Library

EXPRESSWAY

The

Art
Gallery

MACQUARIE

Domain

FM

The Royal Botanic Gardens

A. Gardens Restaurant & Kiosk
B. Pyramid Glass House
C. Rose Garden
D. Grass Plots
E. Outdoor Exhibition
   Starting From Scratch
F. Succulent Garden
G. Visitor Centre & Gardens Shop

• F Toilets - Female
• M Toilets - Male

GATES
1 Palace Garden
2 Shakespeare
3 Woolloomooloo
4 Henry Lawson
5 Victoria Lodge
6 Yurong
7 Queen Elizabeth II
8 Opera House
9 Inner Domain Depot
10 Government House
11 Conservatorium
12 Rose Garden

0   200 m

The Garden Palace (*National Library*)

A few hundred yards from the head of the Cove, towards the left, stands the Governor's house, with its beautiful domain in front ... the whole occupying a space between the head of the cove to Bennilong's Point ... But the domain, beautiful as it is, has lost much of its attraction since being deprived of the kangaroos and emus seen, in Governor Macquarie's time, hopping and frisking playfully about, which never failed to strike powerfully the eye of the stranger on his first sight of them from ship-board. (P. Cunnigham, 1827 from S. Mourot, *This Was Sydney: A pictorial history from 1788 to the present time*)

The kangaroos and emus are sadly gone but this lovely open space has been preserved. Of course, over the years, development of the growing city has nibbled at its edges. The most celebrated was the construction in 1879 of the **Garden Palace**. Many of Sydney's residents disapproved of it at the time, believing it robbed the city of parklands and obstructed the harbour views. Built for Sydney's International Exhibition, the Garden Palace was an enormous structure, modelled after the Crystal Palace in London. The crucifix floor plan measured 250 metres by 150 metres, with the central dome rising nearly 28 metres from the ground. Unlike the Crystal Palace, the Garden Palace was constructed mostly of timber and corrugated iron cladding. After the International Exhibition the building was converted to great halls for concerts, and the galleries were used as museums. Then, in 1882, only three years after completion, it was completely gutted by fire. Many valuable collections, paintings, and archival records were lost. Arson was suspected at the time. All that remains of the great enterprise is a plaque marking the site of the dome, and the Garden Palace Gates in the south-west corner of the gardens.

In 1962 the Eastern Distributor (Cahill Expressway) was completed, ruining the spaces on the south side of the Gardens and providing a barrier between it and the Domain. It should have been tunnelled but even that small bit of tunnelling that saved Macquarie Street from being cut off the Gardens was achieved only after strong and persistent protest from the State's chief planner, Nigel Ashton. Despite a similarly staunch defence in 1987, the Minister for Public Works in the Labor Government, Laurie Brereton, pushed through Cabinet proposals for the Harbour Tunnel and its entrance in the Domain, once more sacrificing Sydney's heritage for unwarranted and destructive roads.

Inside the Gardens

A **Visitors' Centre** and **Gardens Shop**, off Mrs Macquarie's Road, is open 10 a.m. to 4 p.m., Monday to Friday, and 12 noon to 5 p.m. on weekends. Phone 231 8125 for information. Free maps and leaflets are available. Guided tours lasting 90 minutes begin from the Centre on Wednesday and Friday at 10 a.m., and Sunday at 1 p.m. Beside the Visitors' Centre is the **National Herbarium**, with an inquiry counter open 9 a.m. to 4 p.m., Monday to Friday, to answer your botanical questions.

The **Gardens Restaurant**, in the middle of the gardens, opposite the Art Gallery, is a very plesant up-market cafeteria with salads and hot foods and is licensed. Opening times are 10.30 a.m. to 4 p.m. weekdays, and 12 noon to 4.30 p.m., weekends.

The **Pyramid Glasshouse**, across from the State Library near the expressway, contains a collection of orchids, palms, and tropical ferns. Opening hours are 9.30 a.m. to 4.30 p.m., November to February, and 9.30 a.m. to 3.30 p.m., March to October.

Around the Domain

The **Art Gallery of New South Wales**, on Art Gallery Road, in the Domain, built in 1885 has a number of interesting collections, especially the Australian impressionists. Check Part 8 for details. Soap-box orators can be heard at the **Speaker's Corner** in the Domain across from the Art Gallery on Sunday afternoon.

Just down the road from the Gallery, overlooking Wooloomooloo Bay, Garden Island and the finger wharf, the Boy Charlton Public Swimming Pool provides a welcome break from city touring. The 50 metre outdoor pool is open 7.30 a.m. to 7.30 p.m., October to November, and 8 a.m. to 5 p.m., May to September. Boy Charlton is not the first pool at this spot. Opened in the 1850s the Robinson's Hot and Cold Sea Water Baths were divided into ladies' baths and gentlemen's baths. The division of this patch of sea water was achieved by mooring a hulk, the Ben Bolt, between the two bathing areas. The deck of the boat was used for change rooms, and coffee and biscuits were served. The **Woolloomooloo Finger Wharf**, which dominates the bay, is probably the largest wharf of its kind remaining in the world today. It has so far survived numerous demolition threats.

Government House, in the north-west section of the Domain (1845), is a large Tudor-style mansion and the home of the Governor of New South Wales. Government House is only open to the public on special occasions.

6   Hyde Park

When Governor Macquarie first fenced this common and declared it a park in 1810, naming it after London's **Hyde Park**, it was 24 hectares (60 acres), mostly of woodlands, and on the edge of the town. Horses were raced on a quarter-mile track

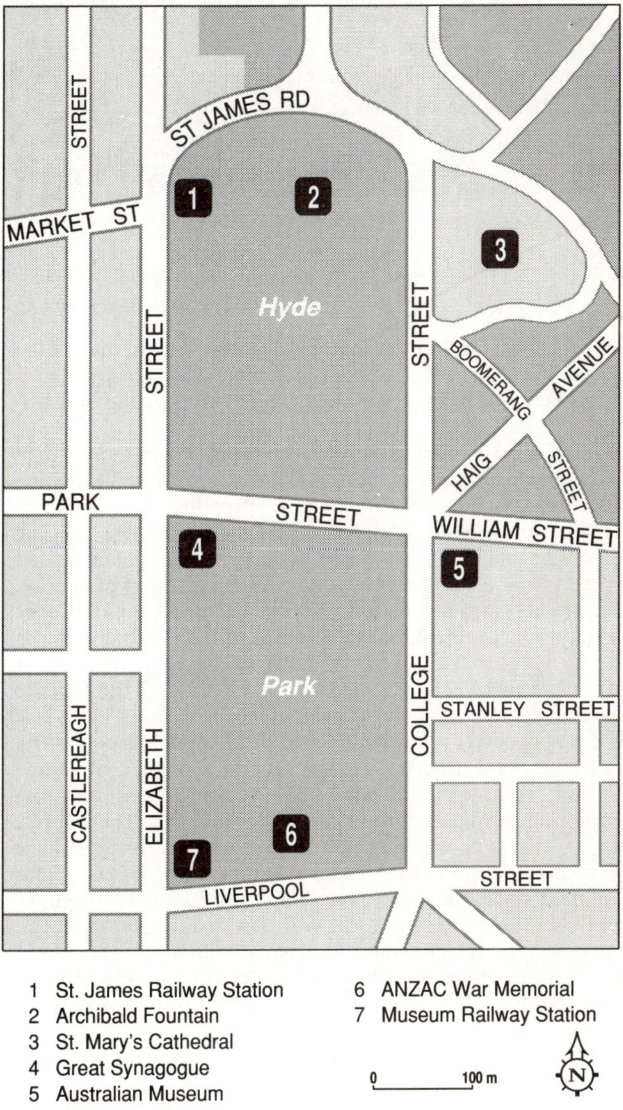

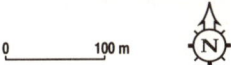

1 St. James Railway Station
2 Archibald Fountain
3 St. Mary's Cathedral
4 Great Synagogue
5 Australian Museum

6 ANZAC War Memorial
7 Museum Railway Station

0   100 m

The Archibald Fountain

until the 1820s, and cricket matches were held until the 1850s. In 1868, an enormous wooden pavilion was erected for a spectacular ball, held in honour of the visiting Duke of Edinburgh.

On sunny days during lunch hours, office workers swarm into the park to eat their lunch. The beautiful walkway the length of the park is shaded by giant, spreading fig trees. It's a great place for a picnic, or just a refuge from the hectic shopping streets. The nearby David Jones department store has a food hall on Market Street with a wonderful assortment of cheese, pates, breads, and salads for a special picnic. A popular spot with visitors is the **Archibald Fountain** at the north end of the park near Macquarie Street. A baroque-like extravaganza, it was donated in 1932 by J.F. Archibald, a founder of the *Bulletin* magazine.

At the southern end of Hyde Park, at Liverpool Street, is the **Anzac Memorial**. Classic art deco, it was designed by Bruce Dellit and completed in 1934. The sculptures around the exterior represent the branches of the services. The central sculpture inside by Raynor Hoff, represents an Anzac soldier upon a shield, supported by his mother, sister, wife, and child. The domed ceiling has 120,000 stars, one for each volunteer who enlisted in the First World War. On Anzac day, April 25, the memorial march ends here with wreath-laying. Inside, there is a moving photographic exhibition on Australians at war: open 10 a.m. to 4 p.m. Monday to Saturday, and 1 p.m. to 4 p.m. on Sundays. Admission is free.

Hyde Park marks the eastern edge of the central business district. On the eastern side of the park, St Mary's Cathedral and the Australian Museum are the most notable buildings. Relentless development along the west edge of the Park, of undistinguished, high-rise office blocks and hotels are ruining the scale of once pleasant streets. They loom threateningly over the Park, casting long shadows, and destroying the views.

The Roman Catholic **St Mary's Cathedral** at College Street and St Mary's Road was completed in 1928 after a construction period spanning more than 50 years. Still not quite complete, the proposed spires on the western side have yet to be made.

The Gothic interior is enhanced by beautiful rose windows. Tours are given of the cathedral on the first Sunday of the month at 2 p.m. The church also has a small museum.

Continuing along College Street you'll see the mammoth **Australian Museum**, 6-8 College Street. Leave plenty of time to explore this important natural history museum. Check details in Part 8.

Behind the war memorial, on Liverpool Street is the **Department of Planning**, 175 Liverpool Street. The Information Centre on the ground floor has for sale, many publications on city planning, and environmental and historical topics. Ask about their excellent pamphlets on walking tours around the city.

The **Great Synagogue** at 189a Elizabeth Street was built in 1878. The entrance is off Castlereagh Street except during services. The lavish interior is well worth a look. Tours and inspection are on Tuesdays, 12 noon to 3 p.m., Thursdays, 1 p.m. to 3 p.m., and Sundays at 10 a.m. to 1 p.m. Phone 267 2477.

On the west side of the Park, on Elizabeth and Liverpool Streets are several good cafes, restaurants, and a few pubs. For sumptuous sandwiches to take away there's "Fed Up", at 281 Elizabeth Street. The "Connaught Deli", at 185 Liverpool Street specialises in bagels and salads and extravagant cakes. Hidden upstairs and around corners in this area are a few Greek restaurants. Check Part 8 for details.

7   Shopping and business district

This section describes the "downtown" of Sydney and takes in the business areas on the north and the shopping areas to the south.

Business district, George Street

The core of the business district falls between Kent Street on the west, Phillip Street on the east, and south of Bridge Street, as far as Martin Place. Wynyard Station is the closest railway station and this area is a short walk from Circular Quay.

George Street (northern end), is Sydney's "main street". Commerce in the colony had its roots in Lower George Street in the Rocks and it quickly spread north. The "Regent Hotel" (1983), 199 George Street, was designed by Michael Dysart, and is perhaps Sydney's top five-star hotel. Grosvenor Place, next to the Regent is architect Harry Seidler's newest high rise completed in 1987 and is 44 storeys tall. Clinging tenaciously to the corner in front is the old **Johnson's Overalls Building**, which is protected by a heritage order, despite Seidler's protests.

In an area of depressing and oppressive anti-human modern architecture, the **Metropolitan Hotel** (1910), corner George and Bridge Street, is like meeting an old friend on a crowded train. During the Federation period a number of lovely pubs were built in Sydney and this is a good example of Federation architecture. Although the interior has been modernised it is still a cosy place. Pubs are becoming a rare and endangered species in central Sydney. In the 1920s there were about 160 and now only about a quarter that number survive. Office and other uses have rendered many of the sites too valuable. The "Metropolitan" has been saved by building over and around it.

The **Australian Stock Exchange**, 20 Bond Street, between George Street and Pitt Street, is not as hectic as New York's and it even closes for lunch. Visitors can watch the trading floor from 10 a.m. to 12.15 p.m. and 2 p.m. to 3.15 p.m., weekdays. Public lectures are normally held at 10 a.m. and 2 p.m. Phone 227 0618 to check if a lecture is on. The lectures last just over an hour and include the history of the Exchange and an explanation of trading. On the other side of Bond Street from the Exchange, is **Australia Square**. Over thirty sites were assembled to make way for this massive but attractive circular building, which was designed by Harry Seidler and completed in 1967. At fifty storeys high, it dwarfed all other buildings. The revolving Summit Restaurant at the top of this building is open for lunch, dinner, and late after-theatre supper.

Wynyard Railway Station has an underground maze of shops leading to the railway platform. This place abounds with fast film developing and fast food. Mrs Field's Cookies and McDonald's can be found down here, as well as rare central city food shops for fish, meat, and bread.

Business district, Hunter Street

The Hunter Street of the late 1800s sounded more interesting than the street today as it was known as the street with the bird stuffers. Two shops belonged to bird stuffers, and the shop windows were filled with deceased parrots, cockatoos, crimson rosellas, parakeets and the like. A little colour is added to this street by the fruit vendors and their barrows that are usually here during business hours.

The **New South Wales Travel Centre** is at 19 Castlereagh Street, near Martin Place. The Centre has an enormous amount of tourism information for both Sydney and the rest of the State,

Martin Place

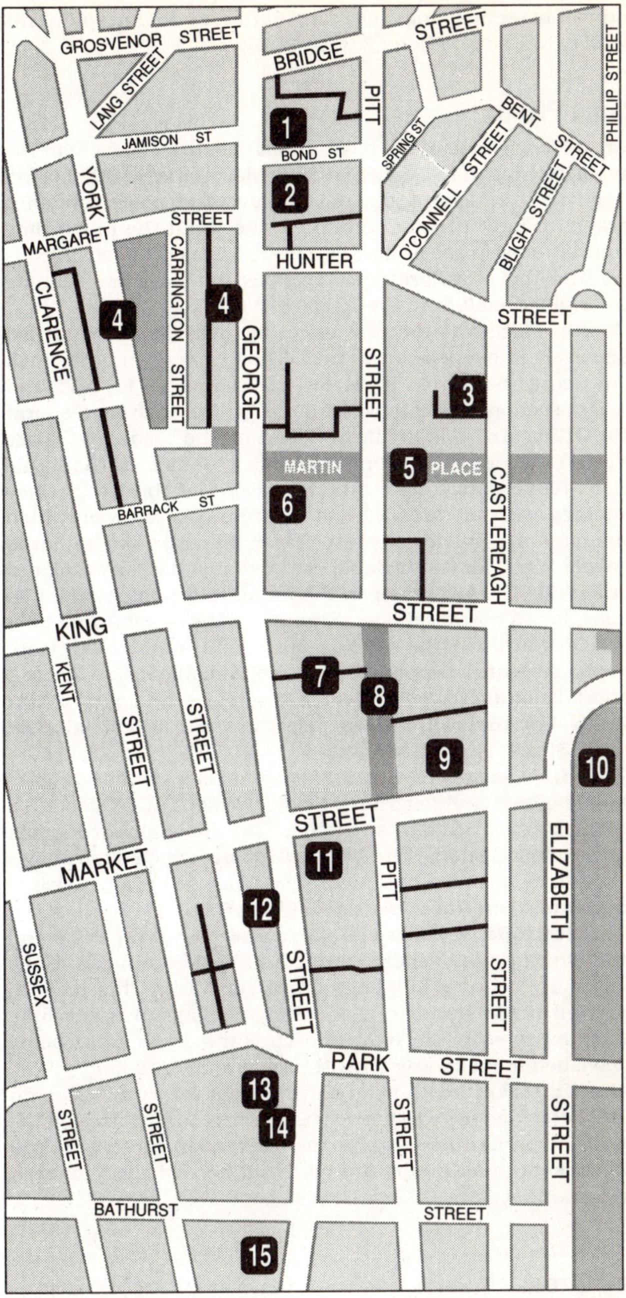

GROSVENOR STREET
LANG STREET
BRIDGE STREET
STREET
PITT
JAMISON ST
PHILIP STREET
1
BOND ST
SPRING ST
BENT
YORK
2
O'CONNELL STREET
BLIGH STREET
STREET
MARGARET
STREET
HUNTER
CLARENCE
CARRINGTON STREET
GEORGE
4
4
STREET
3
STREET
MARTIN
5 PLACE
BARRACK ST
6
CASTLEREAGH
KING
STREET
KENT
STREET
7
STREET
8
9
10
MARKET
STREET
ELIZABETH
11
STREET
PITT
12
STREET
STREET
SUSSEX
STREET
PARK STREET
13
STREET
STREET
14
STREET
BATHURST
STREET
15

1 Australian Stock Exchange
2 Australia Square
3 NSW Travel Information Centre
4 Wynyard Railway Station
5 Martin Place Plaza
6 General Post Office
7 Strand Arcade
8 Pitt Street Mall
9 Centrepoint Tower

10 St. James Railway Station
11 State Theatre
12 Queen Victoria Building
13 Town Hall
14 Town Hall Railway Station
15 Cinema Strip

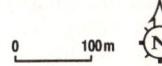

0   100 m   N

and will make bookings. Opening hours are 9 a.m. to 5 p.m., Monday to Friday. Phone 231 4444.

Martin Place

Martin Place is one of the most pleasant "urban spaces" in the city. Stretching from George Street to Macquarie Street, it is full of activity throughout the day. A favourite place for office workers to have their sandwich, or check out the lunch time entertainment in the amphitheatre. The General Post Office is here and the seats are often occupied by travellers reading long-awaited mail from the "poste restante".

Martin Place was the first street in Australia to be turned exclusively into a pedestrian mall. The first section of the mall from George Street to Pitt Street was completed in 1971 and the last section was completed in 1979. The **Sydney General Post Office** (1874) is in Martin Place, on the corner of George Street. Designed by Colonial Architect James Barnet, this Italian Renaissance style building is one of the few Sydney buildings with an arcade. The beautiful clock tower which dominated the Sydney cityscape for many years was removed in World War II for fear of Japanese bombing. It wasn't replaced until 1964. The GPO is open 8.15 a.m. to 5 p.m., Monday to Friday, for all postal services. Stamps can be bought until 5.30 p.m. On Saturday the GPO is open 8.15 a.m. to 12 noon. Telephones (with telephone books) are available inside. There's a special counter for **poste restante** services, and often a long line-up. For **philatelic sales**, use the separate entrance off George Street. Phone 230 7122.

The **MLC Centre** (1978), 19-35 Martin Place on the corner of Castlereagh, is another Harry Seidler design. At 68 floors it's the tallest office building in Sydney. The shopping arcade has many exclusive shops. The **Dendy Cinema**, which often shows foreign films, is in the basement. The "Theatre Royal" is also in this complex, on the south side off King Street.

The **State Bank Centre** (1985), corner of Martin Place and Elizabeth Street by Peddle Thorpe and Walker. Amidst all the turn of the century buildings on Martin Place, this modern structure really stands out. Conservationists unsuccessfully attempted to stop the old art deco Rural Bank from being demolished to make way for this pink granite building.

Halftix Ticket booth is on Martin Place, between Elizabeth and Castlereagh. Buy half-price tickets here for theatres on the day of the performance, no phone reservations accepted. Full price tickets to all events are also sold here at the **Ticketek** window.

York Street

On the west side of George Street, the York Street area has an interesting collection of buildings. The **Grace Building** (1930) 77 York Street, is a Gothic building straight off the streets of Chicago in the 1930s. Originally a warehouse and offices of that retail giant, Grace Bros, it was used for as the Australian headquarters of General MacArthur during World War II. The **AWA Building** (circa 1938), at 45 York Street, is one of Sydney's few remaining art deco buildings, designed by Morrow and Gordon with Robertson, Marks and McCredie. The radio mast on top was designed as part of the building and with the tower this building was the tallest in Sydney until the 1960s. The mast could be seen from far away, even on the other side of the harbour in North Sydney. A display of flashing lights on the mast for many years was used to signal the weather forecast.

Wynyard Park, York Street, behind Wynyard Station, is a forlorn, mostly sunless patch of grass and paving, which has seen better times. Now it's popular with people with bottles in brown paper bags. Public buses depart from here for the North Shore suburbs and the northern beaches. On the south side of the Park is the **Assembly Hall & Scots Church** (1930), York Street, entrance off Margaret Street, which replaced the first Scots Church built in 1826 and demolished to make way for the approaches to the Harbour Bridge in the 1920s.

St Philip's Church (1856), York and Jamison is a lovely sandstone Anglican church designed by Edmund Blacket. **St Patrick's Church** (1844), Grosvenor Street is the oldest Catholic church in Sydney, designed by John Hilly. Until Catholics were given permission to establish in Sydney, they assembled "illegally" on Sundays at a cottage on this site.

Pitt Street

The heart of the shopping area lies between George, Elizabeth, King and Park Streets. Although many large shopping malls have sprung up in the suburbs in recent years, the downtown shopping area still remains vibrant and offers wide choice. Pitt Street is the centre of the retailing area and has been an important shopping street since the early part of the 19th century.

One extraordinary activity reputed to have taken place on Pitt near King Street is rat fighting in the "rat pit". Morris Asher, writing for the *Sydney Mail* in 1907, recalled his early days in Sydney in 1838:

> I saw a dog kill 60 rats in one minute. Horse-racing, dog-fighting, cock-fighting, rat-fighting and prize-fighting were all the rage. Scarcely anything else was thought of, and it was very hard for a young fellow to settle down.

Buskers are the only form of entertainment on Pitt Street Mall

The Strand Arcade

these days. The northern end of Pitt Street was turned into a mall in 1987 after years of opposition from shop owners. There are lots of choices for eating in this area. The "Mid City Centre" has fast food outlets with tables in the middle. "Sky Gardens" also has restaurants on the top floor. David Jones' department store offers an interesting spot for lunch in their Market Street Food Hall. In the middle of a fantastic display of food for sale, is an oyster bar and coffee bar.

The major tourist sight in the area is without any doubt, the wonderful **Strand Arcade** off the Pitt Street Mall, leading to George Street. This lovely arcade was built in 1891 but was completely gutted by fire in 1976. Beautifully restored, it is now home for many boutiques, specialty shops, and charming cafes. Among so many attractive shops, **The Harris Coffee Shop** deserves special mention. Enjoy your choice of fine coffees, sit on a high stool and watch the crowds go by. Internationally known dress designers, such as Maggie Shepperd and Jenny Kee have shops in this Arcade.

Market Street

The **Centrepoint Tower**, on the corner of Pitt and Market Streets, is Australia's largest structure at 305 metres. Two revolving restaurants at the top offer a choice, one expensive, the other a reasonably priced buffet. The view of course is spectacular—a particularly good time to admire the view is around sunset to see both the city in the day and the glittering night lights. Open 9.30 a.m. to 9.30 p.m., Monday to Saturday, 10.30 a.m. to 6.30 p.m., on Sunday. Phone 231 6222.

"Centrepoint Shopping Centre", under the Centrepoint tower is a maze of about 200 shops on four levels. The higher the level, the more expensive the shops. David Jones' (Elizabeth Street Store, and Market Street Store), is Sydney's longest established store, in business for over 150 years. Although a large eight-storey department store, the interior is very elegant, complete with carpets, marble and chandeliers. There's even a doorman at the Elizabeth Street entrance. Grace Bros, the other department store giant in Sydney, is across the street from David Jones, and also has a wide range of goods and services.

The **State Theatre** (1929), at 49 Market Street, is a movie theatre worth a special diversion. Designed by Henry E. White, the term "film palace" comes to life at the State Theatre. Once described as the "Empire's greatest theatre", the souvenir booklet from 1929 described one of its rooms as "having all the lavish splendour and delicacy of...the boudoir of the famous mistress of French royalty".

The auditorium is lit by a 20,000 piece chandelier and the Wurlitzer organ still plays on. The annual Sydney Film Festival is held here.

George Street

The **Marble Bar** of the Hilton Hotel is opposite the Town Hall, on George Street. The Hilton, with its main entrance at 259 Pitt Street, surely must be one of Sydney's ugliest buildings. Inside though, the lovely Marble Bar survives from a High Victorian hotel demolished in 1969. The Italian Renaissance decor of hunting scenes, flowers and fruit in marble, bronze, and ceramic tiles, is wonderfully "over the top". The wall panels were painted by well-known artist Julian Ashton.

Dymock's Book Arcade at 424 George Street is one of Sydney's best and largest bookshops. It's open on Sundays too, and a cafe on the second floor allows a convenient stop among the books.

The **Queen Victoria Building** (QVB) at George and Market Streets, is one of the great success stories of Sydney. Covering a whole city block when constructed in 1898 as a market building, marvelling observers described the QVB as a "Byzantine palace of commerce". Unhappily, it didn't live up to the high expectations. When the produce markets were relocated south of what is now Chinatown in 1910, the QVB ceased trading. Unbelievably, no suitable use was found; for a period it even became a car park. Demolition of the building was seriously considered in 1959. One proposal suggested the transformation of the site into a civic square. Finally, in 1986, the QVB once again opened its doors as a grand shopping centre, with over 200 shops. The transformation is extraordinary and breathtaking. Imaginative and careful renovations retained the turn of the century charm, complete with stained glass windows and a collection of Royal paintings. Now a major attraction for visitors and residents alike, it is open every day. Here is shopping with style. Souvenirs, clothes, bookshops and specialty goods are crowded onto three airy floors. On the ground floor are some classy fast food outlets.

The **Gresham Hotel** (1884), York Street across from The Queen Victoria Building. The facade and the grand staircase inside this pleasant pub are typical of this "late Victorian boom style".

The **Town Hall** (1869), at George and Park Streets, just south of the QVB, has a gracious Victorian exterior, and an impressive interior. Besides being used for council meetings, the Town Hall is renowned for rich and warm acoustics making it perfect for concerts. One of Sydney's early cemeteries was located on this site and the headstones and remains were moved to other cemeteries. But not all of them apparently, for when electric light cables were laid in 1904, a remaining coffin was exposed. A bottle with a newspaper of the day was placed in the coffin which was then cemented over.

On George Street south of Bathurst you'll find Sydney's **cinema strip**. Here are three multi-screen cinemas, fast food chains, video game parlours, restaurants and so on. Most cinemas offer half price admission on Tuesday nights.

Sydney Town Hall

Liverpool Street

Liverpool Street, near George Street, has several restaurants and cafes. At the east end of the street is a pedestrian walkway to Darling Harbour. Between George and Kent Streets are a number of Spanish restaurants. The "Grand Taverna" behind the "Sir John Young" corner pub has great paella and other fish dishes. Jugs of sangria are available at the bar. For a quick lunch or a delicious hot chocolate, check the Cafeteria Espana at 79 Liverpool Street. Adventurous eating includes tripe in tomato sauce, and hearty spanish omelettes.

The "Wilderness Shop" at 92 Liverpool Street raises money to save Australia's remaining wilderness. Come here for great Aussie gifts and information on the Wilderness Society's causes and bushwalks.

Central Railway Station area

The imposing clock tower of the Central Railway station can be seen from a long way. Until the 1930s this end of town was the hub of the retail area, with several major department stores within a short walk from Central. Then the opening of the new train stations on the City Circle helped to pull the shopping district north. After years of neglect, a renaissance is promised. (Although development has been postponed, the proposal is for the gigantic "World Square" project—incorporating hotels, of-fices, and shops—to replace the famous "Anthony Horderns" department store on George and Goulburn Streets.) Hay Street and George Street are also changing very quickly, with Chinatown spreading across George Street.

Central Railway Station remains one of the most important stations on the suburban rail network. Come here for all trains leaving Sydney for the country and other states. The **Traveller's Aid** in the country train station has showers with towel for $3, and $1.50 without towel. A free wheelchair service is available here for disabled passengers (phone 211 2275), and there is also a baby changing room. They have information on places to stay, including a list of backpackers' accommodation. Opening hours are from 7 a.m. to 5 p.m., Monday to Friday, and 7 a.m. to 12 noon on Saturday. Coin lockers are available at the country trian station but must be cleared daily by 11 p.m.

The station has undergone major renovations in recent years, and on the whole has been greatly improved. One less popular renovation, is the removal of the mechanical destination board, replaced by a computerised screen. The destinations were on metal plates that were flipped over by a station hand with a long metal rod. A new home was found in the Powerhouse Museum and now the train timetable is permanently set on a typical Sunday afternoon in 1937.

The **Mortuary Railway Station Terminal** (1869), in Regent Street, just south of the railway station was used by mourners going out to Rookwood cemetery. This unusual gothic sandstone building combines both church and state functions and was recently renovated. It is now used for trains going on special excursions, and VIP trains.

8   Chinatown

Chinatown is to the south-west of the business district and near the southern corner of Darling Harbour. Take the series 400 buses on George Street, or walk 10 minutes from Town Hall or Central Stations.

Dixon Street, Chinatown

Dixon Street is the centre of Chinatown. Colourful Chinese ceremonial archways mark the entrance at each end of this pedestrian street. Chinatown is one of the best places for a meal, however much you want to pay. Sydney's largest entertainment complex for rock concerts and the like, is on the western edge of Chinatown. Whenever you come, you can find plenty of activity in Chinatown, with people dining late every night.

The first Chinese came to Australia during the gold rush. Chinese coming to Sydney settled predominantly in the Rocks and also around Botany Bay where they had market gardens. Many Chinese established shops in the present area when Paddy's market, a large fruit and vegetable market on Hay Street, was opened around the turn of the century. Sadly, Paddy's market was pushed out in 1987 when the site was redeveloped but Chinatown is thriving and continuing to spread into surrounding streets. Unlike San Francisco's Chinatown, not many Chinese actually live in Chinatown. But they flood in from the suburbs, especially on the weekends to shop and dine. Chinese New Year usually falls in late-January early February. There's a small but very colourful parade with paper dragons and fireworks.

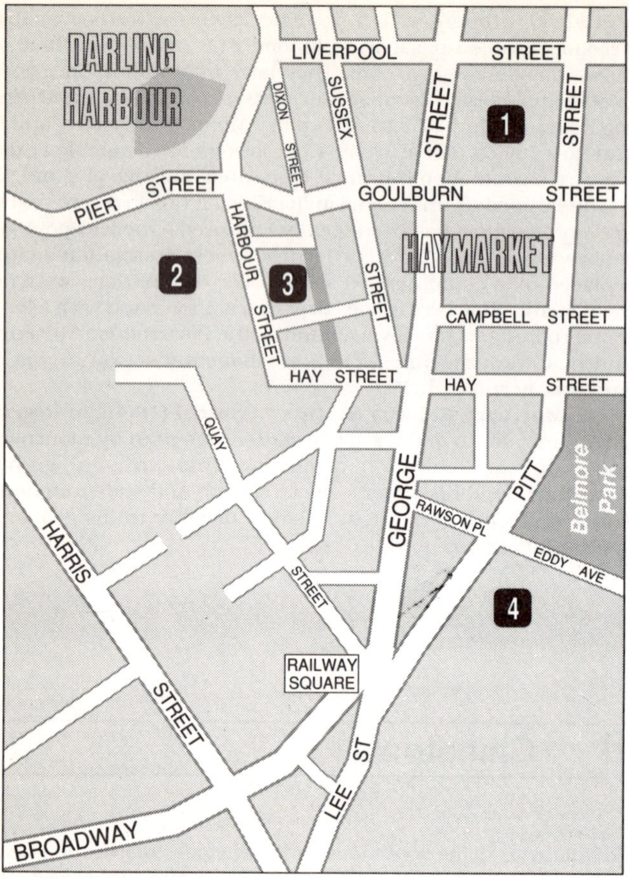

Chinatown

1   World Square
2   Sydney Entertainment Centre
3   Chinatown
4   Central Railway Station

0 _____ 200 m

Food in Chinatown

A few Singapore-style food centres have sprung up in recent years. Look for them in the basement of the "Chinatown Centre" on the west side of Dixon Street, the "Dixon Gourmet Food Centre" on the corner of Little Hay and Dixon in the basement, and the top floor of the "Pailou Plaza" on Dixon Street. Take a good look around before deciding where to eat, there's so much to choose from. You can eat very well for $5. Most types of eastern cooking are featured, including Indonesian, Thai, Vietnamese, Japanese, Malaysian, and of course Chinese. Don't come here for a quiet meal, but for great value, and good quick food. The food centres are open from 10 a.m. to 9 p.m., every day.

Yum cha, sometimes called **dim sum**, has become a popular institution. This is "brunch" time eating in Chinatown. On weekends the more popular restaurants have long queues so book ahead or get there early. Yum cha is served by waiters and waitresses whisking by your table with trolleys full of food. As the trolleys go by, make your choice amongst steamed barbecue pork buns, meat and vegetable filled dumplings, spring rolls, chicken's feet, wonton skins, noodle rolls, custard tarts, and much more. The meal is served with tea; if you would like more tea, simply leave the lid off the teapot and it will be quickly refilled. A yum cha meal costs around $10 in most places, depending of course on your appetite.

The many Chinese shops are certainly worth a look. You may find some bargains among the clothes on offer, and the exotic foods make for an interesting browse. Another delight of Chinatown are the bakeries, especially "Maxims" on the corner of Goulburn and Sussex Streets. Find a huge variety of sweets and savouries such as barbecued pork buns, lotus seed cakes, bean paste buns, dumplings, and walnut cookies.

9   Darling Harbour

Darling Harbour is walking distance from the centre of the city, but can be awkward to get to unless you are sure of the way. Nicest way to travel is by ferry from Circular Quay, but unless you happen to be already at the Quay, this is unlikely to be convenient. The 400 series buses along George Street will take you nearby, or it's a 10-minute walk from Central and Town Hall. There is also the monorail, $2 for the short ride. Large car parks are located on the south-west side of Darling Harbour, off Harris Street.

Darling Harbour is a new recreation and entertainment complex, situated on the west side of the business area. An impressive selection of attractions includes convention and entertainment centres, a marvellous aquarium, Chinese Gardens, and the National Maritime Museum. The Powerhouse Museum is on the western edge of Darling Harbour area. Darling Harbour is the big time for fast food, an enormous choice of kiosks, restaurants, delis and bakeries. And of course, a good selection of souvenir shops.

Darling Harbour takes its name from General Ralph Darling, who succeeded Governor Macquarie in 1821. Since 1984 it has been transformed from an under-used shipping and storage area to a giant entertainment complex fit to win the 1988 State election. Ironically, the scheme backfired: widespread resentment at the high-handed and authoritarian government attitude, and the high costs at a time of public spending cutbacks, was one factor leading to the landslide win for the Liberal opposition in 1988.

The **monorail** was an extremely controversial part of the Darling Harbour project. Opponents argued it would ruin the streetscape of Sydney's streets, and the public transport system would not benefit from such a scheme. Well, it has ruined the streetscape, and the high cost and low capacity of the monorail is bound to lead to greater problems as Darling Harbour develops. The fare was doubled to $2 in its second year of operation.

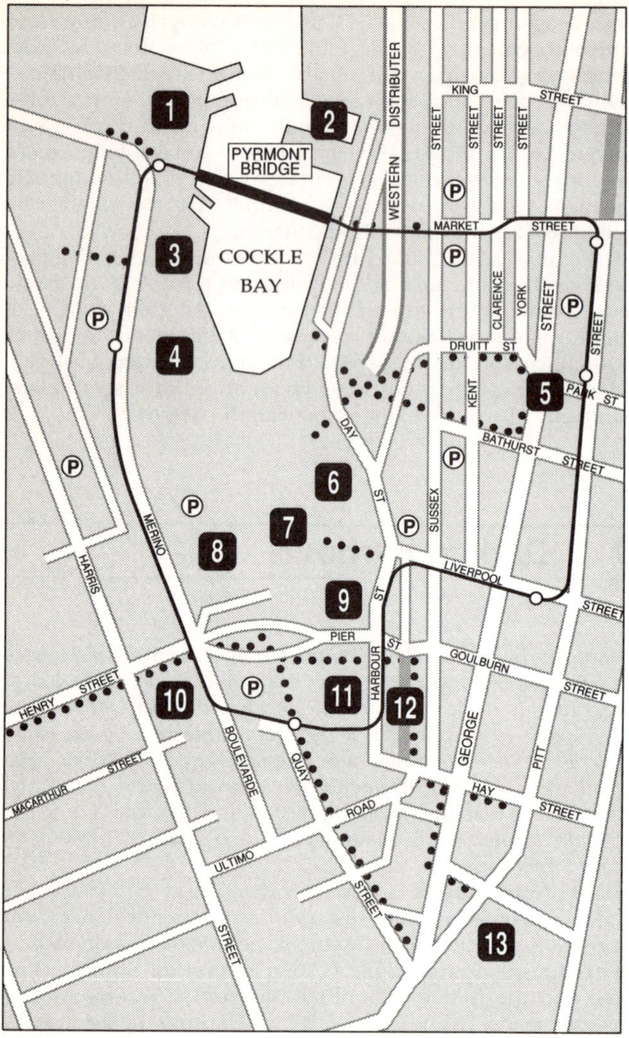

PYRMONT
BRIDGE

COCKLE
BAY

1 National Maritime Museum

2 Sydney Aquarium

3 Harbourside Festival Marketplace

4 Sydney Convention Centre

5 Town Hall Railway Station

6 Darling Walk

7 Tumbalong Park

8 Exhibition Centre

9 Chinese Garden

10 Powerhouse Museum

11 Entertainment Centre

12 Chinatown

13 Central Railway Station

Darling Harbour

1 National Maritime Museum
2 Sydney Aquarium
3 Harbourside Festival Marketplace
4 Sydney Convention Centre
5 Town Hall Railway Station
6 Darling Walk
7 Tumbalong Park

8 Exhibition Centre
9 Chinese Garden
10 Powerhouse Museum
11 Entertainment Centre
12 Chinatown
13 Central Railway Station

Ⓟ   Parking
•••••   Pedestrian Access
—○—   Monorail

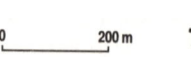

0   200 m

Harbourside (*Darling Harbour Authority*)

Main attractions

The **Sydney Aquarium** is close to the eastern entrance to Darling Harbour. Designed by Philip Cox, the Aquarium is shaped like a giant breaking wave. Among its most popular attractions are the oceanaria. Visitors can walk along the bottom of Sydney Harbour or the open ocean through acrylic tubes in enormous tanks. Appropriately, there is a fish and chip shop here as well as a good gift shop. See Section 6 for admission and hours.

The **Chinese Gardens** were designed by Chinese landscape architects as a Bicentennial gift from the government of Goungzou Province in the People's Republic of China. The atmosphere is one of peace and tranquility, a sanctuary from the glitter and cement of Darling Harbour. If you chance upon a moment of quiet, sit in a pavilion and look over the streams, waterfalls, trees, and flowers, or take tea in the tea room overlooking the lily pond. Admission is $2 for adults, 50 cents children and concessions.

The **Pumphouse Tavern and Boutique Brewery** is located in what used to be the pump house which provided hydraulic pressure to operate lift and bank vaults in the city. It is a delightful pub with a lovely inner courtyard as well as outdoor seating in front. Beer is made on the premises, and you can watch the operations through the courtyard windows. Meals are also available. The Pumphouse is open 10 a.m. to midnight, Monday to Saturday, from 12 noon to 10 p.m. on Sunday.

All the food outlets and shops are found in the **Harbourside Festival Marketplace** on the west side of Darling Harbour. They are open from 10 a.m. to 9 p.m., Monday to Saturday, and from 10 a.m. to 6 p.m. on Sunday. Watch them make meat pies or chocolate eclairs at Harry's, serve yourself in the delightful Yamazaki Japanese bakery, have a steak and onion sandwich at the Cha Cha Grill or treat yourself to a meal in one of the many waterfront restaurants. The 200-plus shops offer fashion, bazaar goods, homewares, and many other things, but provide especially fertile grounds for souvenir hunters.

Along the wharf, beyond Harbourside, is an interesting col-

Darling Harbour, Chinese Gardens

lection of boats. The **James Craig** is a sailing ship built in 1874 and is presently being restored. The **Australian National Maritime Museum** is just beyond this collection of boats. The architect, Philip Cox, wanted to convey the feeling of visiting a shipyard rather than a stuffy museum. Some of the museum's vast collection is appropriately on the water floating in front of the museum. See Part 8 for hours and admission.

PART IV
Around the Suburbs

Once you've explored the centre of the city, you may want to venture into some of the suburbs. The areas described here will give you a good tour of the most interesting suburbs. They contain more of the real Sydney, and they are easily accessible by public transport.

10   Kings Cross

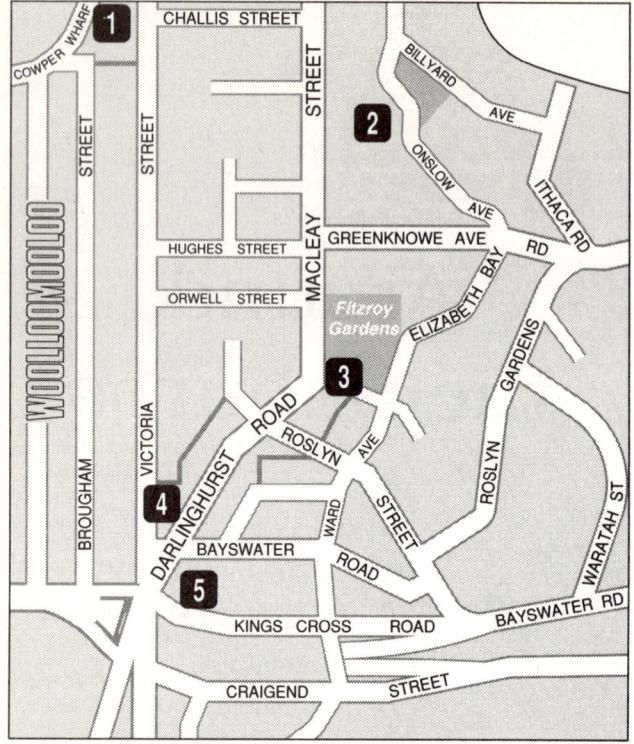

1   Harry's "Cafe de Wheels"
2   Elizabeth Bay House
3   El Alamein Fountain
4   Kings Cross Railway Station
5   Hyatt Kingsgate

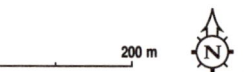

0   200 m

The quickest and easiest way to get to the "Cross", is to take a train on the Eastern Suburbs line to Kings Cross station. Bus numbers 324, 325 and 327 leave the city from Circular Quay and travel via William Street. An attractive alternative from the Quay is bus 311 which travels via Potts Point. From Railway Square, a bus 311 goes via Taylor Square. The walk along William Street from Hyde Park to the Cross only takes about 15 to 20 minutes. A giant Coca Cola sign marks the Cross.

The Cross is best known as the red light capital of Australia. Some of the raciness seems to have left the Cross recently, together with many of the hookers, who have headed down the hill to William Street and Darlinghurst and to more comfortable accommodation in the suburbs. However, escort agencies, X-rated videos, adult bookshops, and topless waitresses are still plentiful. But the the Cross is also an important tourist centre and remains a thriving community. Attempts to reclaim the Cross for the community have sanitised it a little. The Cross definitely has two moods. In the daytime it's a pleasant place for a coffee or a browse around the shops. At night it is decidedly sleazy.

Shifts in character are part of the history and evolution of

Kings Cross

the Cross. In the roaring 1920s, this was Sydney gangland territory and home of the underworld. Chicago-style gang shoot-outs enlivened the atmosphere in the night clubs. Tough drug laws, stiff sentences, and the Depression brought this era to an end. In the 1930s and 1940s, the Cross turned bohemian at a time when the rest of Sydney was very conservative.

Gradually, the bohemian spirit declined giving way once more to its red-light character. In the late 1960s the Cross became a rest and recreation stop for American soldiers during the Vietnam War.

Warning: Be careful any time, especially at night. People do get mugged here.

Darlinghurst Road

If you arrive by train, you will find yourself on Darlinghurst Road, the centre of Kings Cross. It has a European atmosphere, with narrow tree-lined streets, delis, fast food and coffee shops, souvenir, and newspaper shops, adult movie houses and bookshops, night clubs and strip joints. The commercial area is on Darlinghurst Road. The sights are the people on the streets, many of them tourists. At the end of Darlinghurst Road is the **El Alamein fountain**.

You will have no difficulty finding a place to eat. Restaurants and coffee shops for all budgets abound. They include the Oz Rock Cafe, billed as "Australia's first Rock Cafe", on the corner of Darlinghurst and William Street, opposite the Hyatt; McDonald's on Darlinghurst Road by the station; a number of Japanese restaurants, and a Korean restaurant; fancy ice-cream parlours, pies, French pastries and pizzas. Also all sorts of cheap eats and take-aways, including pizza, Lebanese and Thai food. Many of the coffee shops are open 24 hours a day. You might like to check out the "Sweetheart Cafe", opposite McDonald's for a touch of Kings Cross nostalgia.

This is also a good place for food shopping, as there is a good selection of delicatessens, a butcher, bakeries and fruit shops and barrows. Buy interstate and foreign newspapers and magazines at the newspaper shop next to the station.

Parking can be a problem, especially at night. A large municipal underground car park is located close to the El Alamein Fountain.

Victoria Street

Victoria Street was a street under siege in the 1970s. The fine, late-Victorian terraces were going to be demolished to make way for more skyscrapers. The unions came to the aid of the passionate resident groups and issued work bans. There are some particularly fine Victorian terraces around the corner on Challis Avenue, numbers 7 to 21.

Macleay Street and Potts Point

At the far end, furthest from the Hyatt, Darlinghurst Road turns the corner by the El Alamein fountain and becomes Macleay Street. This is the quieter, more charming suburb of Potts Point. Away from the hype of the Cross, life is less hectic, the restaurants and hotels up-market. If you want to stay near, but not in the Cross, this is a good choice. The further you head down Macleay Street, the quieter and more refined it becomes. At the Cross end, there are still many backpackers hostels.

The famous **Wayside Chapel** and drop-in centre is here, at 29 Hughes Street. Established by Ted Noffs to help the down-and-out, this inter-denominational chapel has provided a valuable community service in the Cross since 1964. There is a 24-hour crisis counselling service, a cafe, and a theatre.

Backpackers' Kings Cross

Since the explosion of travellers visiting Australia, Kings Cross has become the centre for accommodation and entertainment. Oddly, after its checkered and lurid history, the Cross is becoming something akin to Earls Court in London. And just like Earls Court, many of the travellers are Australian. Backpackers' hotels and hostels seems to be everywhere, but Darlinghurst Road, Victoria Street and around Springfield and Orwell Streets are the current favourites. Many of the hostels have noticeboards. If you're on the look out for a ride to Cairns, or a Holden to drive the Nullarbor, you might like to test the market here on Victoria Street.

Entertainment

Kings Cross is well known for its entertainment. There is a good choice in nightclubs and strip joints, not to mention the famous female impersonators at "Les Girls". At the "All Nations" the best of Sydney's musicians often improvise until the small hours of the morning. The Oz Rock Cafe has live entertainment, and there is a choice of gambling clubs on Darlinghurst, Kellet, and Bayswater Roads.

Elizabeth Bay House

On the edge of Kings Cross and an easy walk from Macleay Street, is Sydney's most famous colonial mansion. Bus 311 from Circular Quay stops at the front door. See Part 8 for hours and admission. This elegant home was designed by John Verge in 1835 for the Colonial Secretary. It is furnished in 1840s furniture and the house has wonderful views over the harbour.

Garden Island

At the end of Macleay Street there is a naval station called Garden Island. It is only open to the public on special occasions. Better views can be had from the harbour or on the other side of Woolloomooloo Bay.

Garden Island marks the place where World War II came to Sydney Harbour. On 31 May 1942, the American cruiser **USS Chicago** was moored at Garden Island when it spotted the periscope of a submarine. It fired and hit its target which turned out to be a Japanese two-man midget submarine. A second midget submarine aimed at the Chicago but missed and hit a floating workshop killing 19 men. In the early hours of the morning the sub was spotted by a patrol boat and was depth-charged. Divers recovered the wreckage and restored portions are now in the Australian War Memorial Museum in Canberra.

The only other incident was on 7 June when submarines again entered the harbour and fired at harbourside homes. This sent harbour properties plummeting and choice flats and homes were sold for a song.

Woolloomooloo

The 'Loo is the residential area east of the Domain and west of Kings Cross, south of William Street. From Victoria Street you can go down the steps to Dowling Street.

Many overseas visitors are amused by this suburb's Aboriginal name, but few can spell or pronounce it. Its meaning is obscure, though two suggestions have been "young kangaroo" or "resting place for the dead". From the mid-19th century this residential suburb has been shared by the gentry along the cliff and by the working class down in the basin. The 'Loo maintained a strong waterside community identity into the 20th century and many of its residents worked at the nearby docks or as fishermen. It has survived as a residential area but not without its battles. In the early 1970s almost all of the 'Loo was included in an enormous redevelopment scheme for office blocks, hotels, and apartment blocks. The unions once again responded to the plea to help "stop knocking down other workers' housing" and issued work bans on the site. In 1975 the Federal Government stepped in and initiated a large public housing project, retaining all the homes in good enough condition to renovate. Traffic was removed from many streets and a very pleasant neighbourhood recreated.

Harry's Cafe de Wheels on Cowper Wharf Road has been churning out pies, peas, and mashed potatoes since 1945 from two trailers; it's open until the wee hours of the morning.

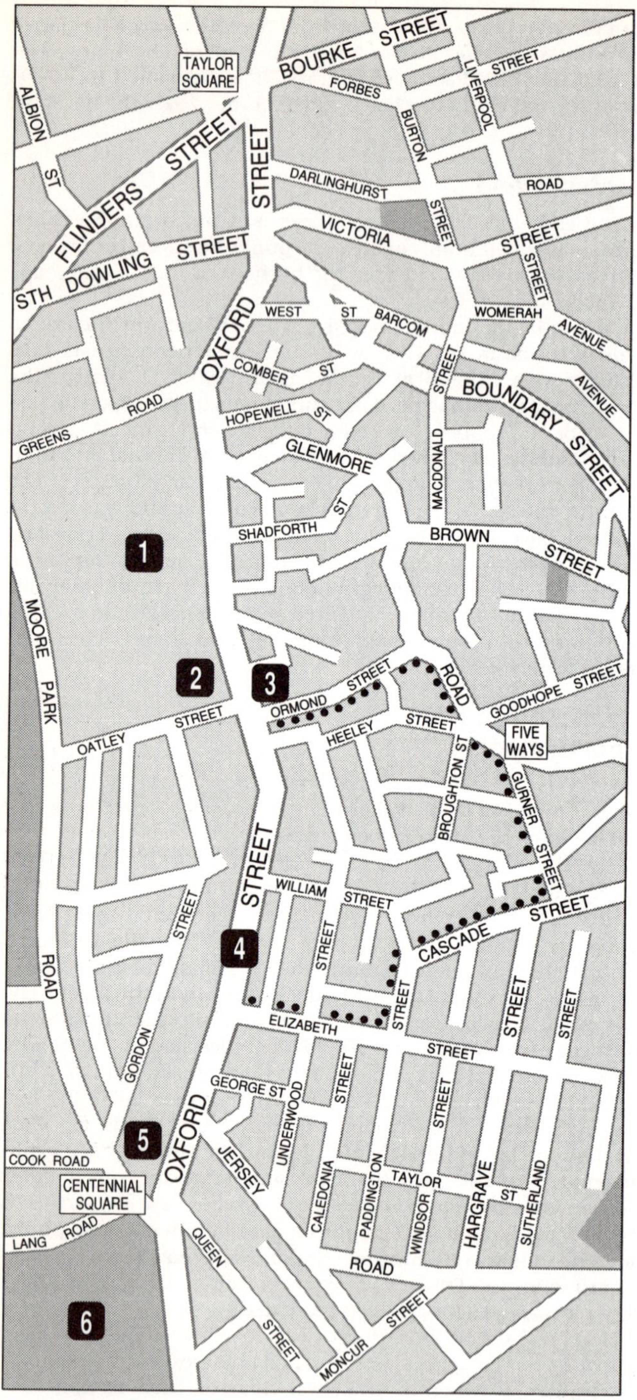

Map labels:

TAYLOR SQUARE

BOURKE STREET

FORBES

BURTON

LIVERPOOL STREET

ALBION ST

FLINDERS STREET

STREET

STREET

DARLINGHURST

VICTORIA

STREET

ROAD

STH DOWLING STREET

OXFORD

WEST ST

BARCOM

WOMERAH AVENUE

STREET

BOUNDARY STREET

AVENUE

COMBER ST

HOPEWELL ST

GLENMORE

GREENS ROAD

MACDONALD

BROWN STREET

STREET

SHADFORTH ST

MOORE PARK

1

2

3

ORMOND STREET

ROAD

GOODHOPE STREET

OATLEY STREET

HEELEY STREET

STREET

FIVE WAYS

BROUGHTON ST

GURNER STREET

WILLIAM STREET

STREET

STREET

CASCADE STREET

4

STREET

ROAD

GORDON STREET

ELIZABETH STREET

STREET

STREET

STREET

GEORGE ST

UNDERWOOD STREET

STREET

STREET

COOK ROAD

5

JERSEY STREET

CALEDONIA

PADDINGTON

TAYLOR STREET

WINDSOR

HARGRAVE ST

SUTHERLAND

OXFORD

CENTENNIAL SQUARE

LANG ROAD

QUEEN STREET

ROAD

6

MONCUR STREET

Legend:

1 Victoria Barracks
2 Paddington Town Hall
3 Juniper Hall
4 Main Shopping Centre
5 Village Bazaar
6 Centennial Park

 •••• Suggested Walking Tour

0 _____ 200 m

N

11   Paddington

Buses 380 and 389 leave from the Quay and travel via Elizabeth Street, to Liverpool Street. Oxford Street begins at the south east corner of Hyde Park. Bus 389 will take you right through Paddington to Fiveways. You may also take the train to Edgecliff railway station and walk through the residential streets to Paddington.

Paddington and Oxford Street lie between Kings Cross and the Eastern Suburbs. Not as racy as Kings Cross, not as sedate as the suburbs, it has an arty feel. **Oxford Street** is a thriving shopping strip over 4 kilometres long. From the busiest, noisiest end at Taylor Square, Oxford Street gradually changes its character, until arriving at Centennial Park, where it transforms into to a quiet residential neighbourhood.

Paddington is famous for its charming terrace houses with iron lace balconies. These are often now lovingly renovated by the prosperous residents. Most of these closely huddled homes were built in the mid to late 1800s. A typical terrace is set back from the street behind an iron palisade fence and has a double-storeyed cast iron verandah. They often have additional ornamentation such as stained glass windows, ceramic tiles and decorative chimney pots.

Although Governor Macquarie had a road built through the Paddington area down to Watsons Bay as early as 1810, not much activity occurred until the Victoria Military Barracks opened in 1848. Soldiers homes were built first, and the shopkeepers followed. From 1860 to 1890 this area was a builder's dream and close to 4,000 houses were built in this period. The terrace design lent itself well to quick development and allowed the land to be carved up into tiny parcels. Many of the builders worked with parcels of land that would accommodate 4 to 6 houses. Often they would build one for themselves at the end of the row, which accounts for the more elaborate corner houses.

Around the turn of the century suburban life in the garden suburbs became popular with the middle class. At this turning point, Paddington began its decline into slum conditions. Many planning proposals called for wholesale demolition. But as early as the 1950s the middle class rediscovered the potential and charm of Paddington and began moving back and renovating.

Walking Oxford Street

If you choose to walk the length of Oxford Street, it's close to 4 kilometres from its beginning at Hyde Park to Centennial Park (from Centennial Park, Oxford Street becomes a residential street, until it ends at the commercial centre of Bondi Junction). There are plenty of shops, cafes, and pubs to tempt you along the way.

Taylor Square is a little like Picadilly Circus, lots of traffic and the appearance of things going on. There are plenty of restaurants here and gay clubs and bars. On the north side is the imposing Darlinghurst Court House. Further east, on the south side at South Dowling Street, is the Academy Twin Cinema, with eclectic taste in movie selection. Still heading east, Oxford Street begins to change its character once more as it turns the corner. On the south side is **Victoria Barracks**, one of the highlights of Paddington. Completed in 1848, replacing the original barracks on George Street, the Victoria Barracks was occupied by 800 British troops until 1870. Still in active use by the army, the Barracks is open to visitors only once a week. On Tuesdays, at 10 a.m., visitors can watch a ceremonial changing of the guard, followed by a tour of the grounds and buildings and a small military museum (see Part 8 for details).

A little diversion off Oxford Street across from the Barracks and down Shadforth Street will take you to the narrow, winding streets of the **original settlement** in Paddington. The earliest cottages are on Gipps and Prospect Streets and were built in the 1840s.

The **main shopping area** starts beyond the Barracks with good souvenir shops, antique shops, book shops, clothes shops, delis, and lots of restaurants and cafes, with supermarkets and fruit shops, among the boutiques. Watch out for Folkways, an excellent music shop and New Editions, a great book shop with a cafe attached.

Historic buildings mark the beginning of the "village" of Paddington. On the corner of Osmond and Oxford Street is the lovely **Paddington Post Office** built in 1885. Across the street is the elegant **Juniper Hall**, probably the oldest surviving villa in Australia. It was built in 1824 by the wealthy distiller, Robert Cooper. His distillery was down near Sutherland Street. The name Juniper comes from the chief ingredient of gin, the juniper berry. In the 1850s Juniper House was converted to an institution and used for a number of purposes over the years. In the 1920s it was converted to an apartment block and its frontage was blocked by a collection of shops. The National Trust has renovated the property and it now houses the **Museum of Childhood** (details in Part 8). A National Trust Gift Shop and a coffee shop are inside. The **Town Hall** is a product of the late 19th-century boom times, completed in 1891 and designed by J.E. Kemp. The **Village Bazaar** has become a Paddington institution on Saturdays. Over 250 stalls display second hand goods, arts and crafts, and even antiques, plus plenty to eat. Design and fashion students often display their creations. Located on the south side of Oxford Street, past the

Paddington terraces

intersection of Elizabeth Street in the Uniting Church grounds, the Bazaar is open on Saturday, 9 a.m. to 4 p.m.

The Finest of the Paddington Terraces

For a good look at some fine terraces you simply have to venture north off Oxford Street into the side streets. It's fun to wander around; you'll always find your way back to Oxford Street.

Ormond Street, near the Paddington Town Hall, is fairly typical. If you continue down Ormond, then right onto Glenmore, you'll soon end up at **Fiveways**. The Fiveways intersection is marked by the attractive Royal Hotel. The village shops are quieter now than when the tram to Bondi passed through here. If you continue up to **Gurner** you'll be rewarded by a sight of particularly fine terraces. Next turn right on **Cascade Street**, left onto Paddington Street and Elizabeth Street will take you back to Oxford Street.

If you find yourself at the other end of Oxford Street (the east end), and haven't seen the terraces yet, go down Jersey Road to Paddington Street and then follow the directions above. Another very pleasant walk at this end of Paddington is to stroll down the elegant shops on Queen Street. "Zigolini's" at 107 Queen Street is a pleasant cafe which draws a regular breakfast crowd. Make sure to take a little diversion down the quaint **Spicer Street** off Queen Street to look at the houses.

Centennial Park

For an escape from urban Sydney, look for the Oxford Street gates leading to Centennial Park. The park was a gift to Sydney from the Government on 100 years of settlement in 1888. Macquarie had set this area aside as a common and the swamps here provided Sydney's first permanent water supply in 1837.

This English-style park covers a massive 220 hectares and has a wonderful variety of landscapes ranging from formal gardens to wooded areas. There are numerous ponds, bike, walking, and riding trails, and it is one of the best birdwatching places in the urban area. On weekends people flood into the park from all over Sydney but there's plenty of room for everyone.

There's a cafe on the corner of Grand and Park Drives open from early morning to dusk. **Cycling** around the park is a very popular activity and there are bicycle hire shops on Clovelly Road and a few places that hire horses. Try to book ahead for the weekends. See Part 8 for details.

12   Glebe and Sydney University

Buses for Glebe leave from George Street, Broadway near Central Station, and Castlereagh Street. Or walk out west of Central Station along Broadway. Several buses go from the city along Parramatta Road (with 400 numbers), and buses 431, 432, 433, 434 go down Glebe Point Road.

The Glebe has its own special village atmosphere. Close to

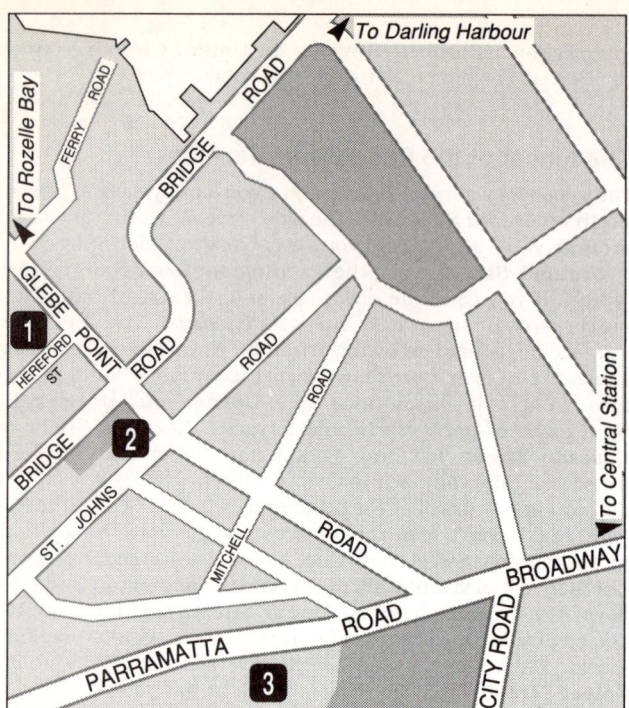

To Darling Harbour

To Rozelle Bay

FERRY ROAD

BRIDGE ROAD

ROAD

GLEBE POINT ROAD

HEREFORD ST

ROAD

ROAD

BRIDGE ROAD

ST. JOHNS

MITCHELL

ROAD

ROAD

BROADWAY

PARRAMATTA ROAD

CITY ROAD

To Central Station

1

2

3

1   Youth Hostel
2   St John's Church
3   University of Sydney

0   300 m

N

Sydney University, many students and staff live in the area, and help to support a great variety of cafes, restaurants, pubs and even the occasional bookshop. Glebe is a good alternative for a place to stay if you don't mind a short bus trip into the city. The leafy streets provide a retreat from the hustle and bustle of the city. Glebe is also a good alternative for backpackers who prefer to avoid the Cross. Three Youth Hostel Association lodges are located in the immediate area. On-campus accommodation is available during the university holidays in July, January and February.

The rich variety of 19th century housing in this neighbourhood, ranges from mansions to tiny worker's cottages. With rising house prices this residential area has become less of a student housing area and more for those seeking a charming house in an inner city location. Many of the large terraces which had been converted to boarding houses, have been reconverted back to a single residence in recent years.

Glebe Point Road

Glebe Point Road is the main street in the Glebe. It's a sure place to find a good bookshop, health food shop, take in a movie or a late night eatery or cafe. On the restaurant front you've got a choice of just about anything, from Russian to Thai and generally at reasonable prices. Most of the restaurants have menus posted outside so you can take a good look around before you decide. One of the things that make Glebe Point Road such an interesting walk is the combination of commercial and

Glebe Point Road shops

residential buildings. Intermingled with the shops are stately
Victorian mansions and run-down cottages with wild, over-
grown gardens.

Glebe Point Road is a long street and the commercial area
extends for some eight or nine blocks. If you have the time, it's
pleasant to walk the whole strip and perhaps take a bus back
or maybe rent a bike from Inner City Cycles at 31 Glebe Point
Road.

A diversion off Glebe Point Road, east onto Mitchell Street,
will take you to the Glebe Point Estate. Following the unpopular
high rise public housing projects in the 1960s, the Commonwealth Government converted this mid-19th century housing
to a public housing estate. The original cottages were very
successfully renovated.

At the end of Glebe Point Road closest to Sydney University
is an interesting mix of shops, cafes, and restaurants. "Lolita's
Cafe" at No. 29, has the best coffee prices and "Badde Manors
Cafe" at No. 37, is open from 8 a.m. to 1 a.m. The "Glebe Ritz"
at No. 39 is another favourite with good food. The "Cafe Troppo",
at 175 Glebe Point Road is famous for its cakes, and interesting
fruit drinks, with magazines to read over coffee but is a bit
expensive. Renowned for Vietnamese food is the "Kim-Van",
147-149 Glebe Point Road. For Russian dining, try "Rasputin's"
at 101 Glebe Point Road and just next door is the very popular
"BJ's" with lots of choice on their international menu. Off Glebe
Point Road, at 58 Cowper, is another "No Names", in the "Friend
in Hand Hotel". As always, No Names has good pasta at
unbeatable prices.

Working your way down towards the other end of Glebe Point
Road you'll pass a couple of excellent bookshops. The
"Cornstalk Bookshop" at No. 112 has a large selection of second
hand Australian books and they're open on Sundays. The "Da
Capo" music store is next door. "Glee Books" is another good
bookshop at No.191, open from 9 a.m. to 9 p.m. every day.

The lovely **St John's Church** on St John's Road was built in
1868 and is worth a visit and a stroll around the grounds. On
the corner of Glebe Point Road and Bridge Street, is a park with
massive fig trees. And right at the end of Glebe Point Road
Jubilee Park is a patch of green with good views of Sydney
harbour.

University of Sydney

The main gates to the university are off Parramatta Road. Ask at the gate for a map and directions. This is Australia's oldest university, established in 1850. It is also one of the largest, with over 20,000 students. The pleasant campus has many fine buildings. The **Main Building** with its Great Hall, is a Gothic revival masterpiece of local Pyrmont stone and was completed in 1857. Professor Leslie Wilkinson, who held the first chair of architecture in 1918, designed the elegant **Physics Building**. There are a couple of museums on campus as well as a theatre. The campus cafetarias and coffee lounges are cheap and open to the general public.

---

13   Balmain

Ferries from the Quay go to Darling Street Wharf. Catch buses 442 and 445 from Darling Point to the centre of Balmain. Alternatively, from Long Nose Point, a short walk to Birchgrove will allow you to meet bus 441. Balmain is also easy to get to by buses 441 and 442, leaving from York Street, behind the Queen Victoria Building.

For an interesting walking tour of Balmain, start at the Darling Street Wharf and walk to the Balmain Town Hall. From here you may walk back to either ferry or catch a bus from here to the city (Buses 441 and 442). Another pleasant way to get a glimpse of Balmain is to go for a stroll before an evening meal at one of the many good restaurants near Balmain Town Hall.

Balmain is one of Sydney's oldest suburbs. Although it has been overrun by renovating yuppies it still retains its charm as a wharfside suburb. It's not just a pretty residential suburb, it still has some remnants of its former illustrious, industrial past lingering on, such as the Colgate factory. Some of the old wharves are used as ferry depots.

Darling Street

Walking Darling Street will give you a good taste of the atmos-phere of Balmain. The last stretch of the former tram route to the Darling Street wharf proved too hard to negotiate. To get up and down this steep hill, a "dummy streetcar" was pushed going down and pulled going up. Running parallel to the dummy car another weight pulled or pushed on a pulley system.

One block from the ferry, take a slight diversion down **Johnston Street** to No. 12, a gracious two-storey residence with an enormous second-storey verandah. Built around 1860, it was originally one storey. One further block west, on the other side of the road, take another diversion down **Duke Street** to the water's edge and the ferry depot. Opposite Duke Street is **Datchett Street**. This is one of the narrowest streets in Bal-main and is lined with interesting houses.

The little stone cottage at 177 Darling Street was built in 1843 and was supposedly built in exchange for a barrel of rum. Next store at No. 179 you'll see the **Balmain Watch House** (1854). A former jail, it is opened by the Balmain Association on Sundays

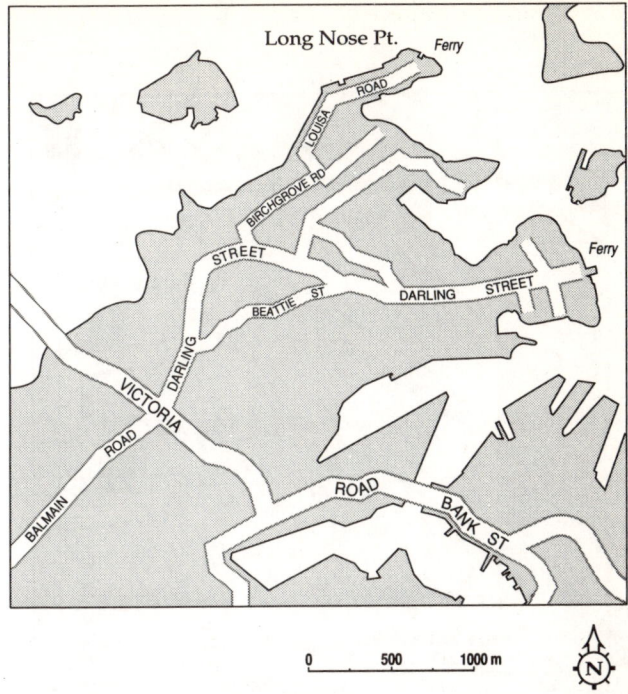

from 2 p.m. to 4 p.m. Excellent self-guided tour maps are available here. A good place to stop for a beer or even a meal is the "London Hotel" at 234 Darling Street.

Balmain Markets are held in St Andrew's churchyard each Saturday, offering all sorts of mostly second-hand things such as records, jewellery, and books. There are good snacks on offer too.

Loyalty Square is a rare sight in Sydney, a lovely square remaining intact with 19th century buildings framing it. The **town hall**, **post office**, and **courthouse** are all grandiose buildings in High Victorian Style, giving one an idea of Balmain as a flourishing commercial centre around the turn of the century. The "Unity Hall Hotel" is a popular hotel with live music. Restaurants and cafes abound in Balmain and several are in the Loyalty Square to Town Hall area.

Walk to Long Nose Point

The walk from Darling Street to Long Nose Point Wharf (the other ferry wharf) is about 2 kilometres. The most interesting way is down Cameron Street to Louisa Road. The **Louisa Road** houses are jammed close together with tiny frontages. These people don't live in totally claustrophobic conditions, however, as most of them have water views out the back and probably a balcony or two to take advantage of them.

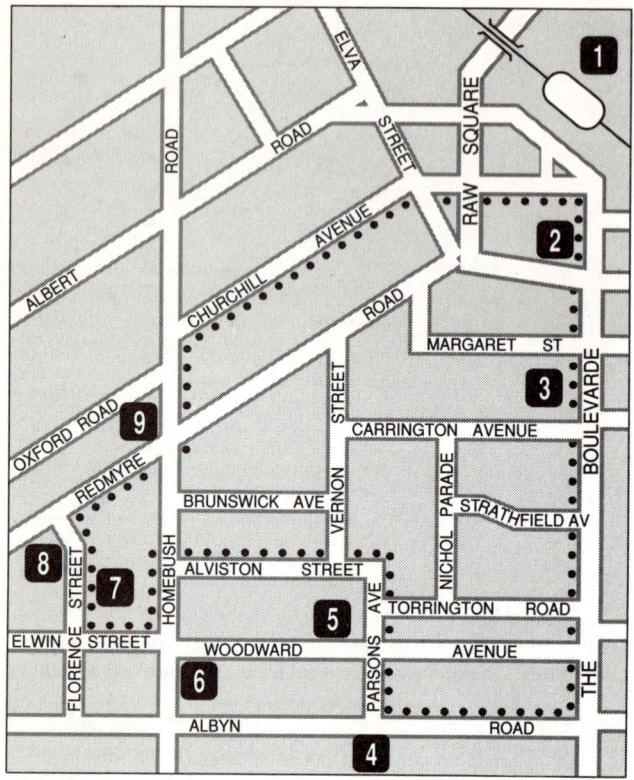

Strathfield walk

1 Strathfield Railway Station
2 Keary's Corner
3 Santa Maria Del Monte
4 "Darenth"
5 Workers Cottages

6 "Halycon"
7 Late Victorian Homes
8 Woodstock Nursing Home
9 Strathfield Council Chambers

Length: Approximately 3.5 kilometres

Source: Strathfield Municipal Council

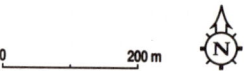

14   Strathfield

The commercial centre of Strathfield is located at the Strathfield Railway Station, about 14 kilometres west of the city centre.

A look around the residential areas of Strathfield gives an idea of an Australian suburb at the turn of the century. Around this time the middle class began to leave the crowded inner suburbs, such as Paddington and The Glebe, in search of larger houses and properties. Suburbia more or less followed the public transport routes. In 1877 a train stop was added at Strathfield on the Sydney-Parramatta line. The beginnings of Strathfield date from this time.

A walk around The Redmire Estate is outlined below. This walk provides an interesting mixture of architecture spanning more than a century from Victorian times to the blocks of flats of the 1960s.

Along this walk you'll see fine examples of Victorian, Federation, and California bungalow-style architecture. The changes made in the two or three decades from Victorian style to the unique Australian Federation style are quite remarkable.

Federation style was characterised by red brick walls and roofs of terra cotta tiles with numerous gables and spires. Other features included porches and verandahs with wooden posts. This style was popular from the 1890s and into the first decade of the 20th century, until shortly after the federation of Australia as a nation.

Strathfield Municipality has prepared walking tours of three of its residential areas and the information below comes from these tours. Pamphlets are available from Strathfield Municipality, 65 Homebush Road, call 76 0431.

The Boulevarde

This walking tour of Redmire Estate begins at Strathfield Railway Station. From the railway station, go south along the Boulevarde. You will come across a collection of shops, many dating from the turn of the century. On the corner of Redmyre Road and The Boulevarde is "Keary's Corner" dating from 1912. Originally a fruit shop, the interior has many period fittings.

49-57 The Boulevarde (1907), is a grand house that remained a private home until 1943 when it was converted into a school. Many of the opulent houses of this period were converted to other uses when maintenance and servant costs became too high. The house at No. 59-63 was converted to a reception and catering establishment as early as 1928 and is now part of Santa Maria Del Monte.

89-93 The Boulevarde, are three single-storey houses built in the California Bungalow style, which was very popular in Sydney in the 1920s. Large projecting verandahs, tiled roofs and shingled gables are typical of this style. Nos. 97-101 The Boulevarde, are late Victorian, built in the early 1890s. Note the bay windows, slate roof and lovely gates and fences.

Albyn Road to Homebush Road

Turn right (west), into Albyn Road from The Boulevarde. At No. 32-34, "Darenth" is a Federation house built at the turn of the

Federation house, Strathfield

century. Turn right into Parsons Avenue then left into Wood-ward Avenue. At No. 5 are two worker's cottages, pretty well in their original condition. Back onto Parsons, and then left onto **Alviston Street**. This pleasant, well-treed street has a mixture of houses from the last century. Turn left onto Homebush Road and you'll see the handsome two-storey Victorian home at No. 81. Continue along Homebush and there are three fine houses, Nos. 89, 91, and 93 built around the turn of the century. "Halcyon", at 110 Homebush Road was built in 1916 and is similar to many of the spacious homes built in Strathfield following World War I.

Elwin Street to Churchill Avenue

From Homebush Road, turn right into Elwin Street and then right into Florence Street. Nos. 14 and 16 are quaint, late Victorian homes. Taking a small diversion left at Redmyre, you'll see an imposing Victorian house, now the "Woodstock Nursing Home", probably built in the mid-1880s. Return to Florence Street and then continue up Redmyre Road to **Strathfield Council Chambers** on the corner of Redmyre and Homebush Roads. The council chambers were designed by the famous architect John Sulman, and completed in 1887. Additions were made in the 1920s. Turn left onto Homebush Road and right onto Churchill Avenue which will bring you back to the railway station. Along Churchill Avenue are several hand-some Federation homes.

15 Parramatta

Parramatta is about 25 kilometres west of the city centre. The train service from the city stations is frequent. If you catch an express, the trip will take about 30 minutes.

Governor Phillip selected the site of Parramatta as a market centre in 1788. Parramatta seemed to have all the ingredients needed for success with the surrounding fertile land and easy transport on the Parramatta River. Yet the settlement grew slowly and was soon overtaken by Sydney. Today Parramatta is located about midway in the metropolitan area. The commercial centre of Sydney is so far away from the Western Suburbs, that now Parramatta is rapidly becoming the major centre for Western Sydney. Office buildings sprung up during the 1980s building boom. Despite all the change, Parramatta has managed to retain some of it fine old buildings as well as the oldest colonial building in Australia.

For a more detailed guide and walking tour, a joint publication by the Department of Planning and the National Trust costs $1 and is available from the Tourist Information Office.

Church Street

This is the main street of Parramatta. It is a short walk down Dacey Street from the railway station. The **Church Street pedestrian mall** was opened in 1987 and has made a big improvement on the atmosphere of the commercial centre. The Parramatta Tourist Information centre is on Market Street just off Church Street, on the north side of the Parramatta River.

St John's Church is at the southern end of the mall. The church occupies the site of the first Anglican church, built around 1800. Nothing remains of the first church but the twin towers date back to 1818-20. The existing church opened in 1855. **St John's cemetery** on the corner of Argyle and O'-Connell, one block west of the church has the graves of many of the First Fleeters. The oldest headstone dates from 1791. The **Centennial Memorial Clock** (1888) in St John's Square makes for unusual street furniture. The **Town Hall** (1883) opposite the church displays the boom times in the late 19th century. Further along Church Street is the handsome former **Post Office** building (1880).

Strathfield Town Hall

George Street, West of Church Street

One block north of St John's Church, just beyond the post office is George Street, Parramatta's original main street. The few remaining 19th century buildings convey the atmosphere of a former stately street. The west end of George Street leads to Parramatta Park and Old Government House while the eastern end leads down towards the river. Phillip's grand vision was for long vistas from Government House to the river, a distance of 1.5 kilometres.

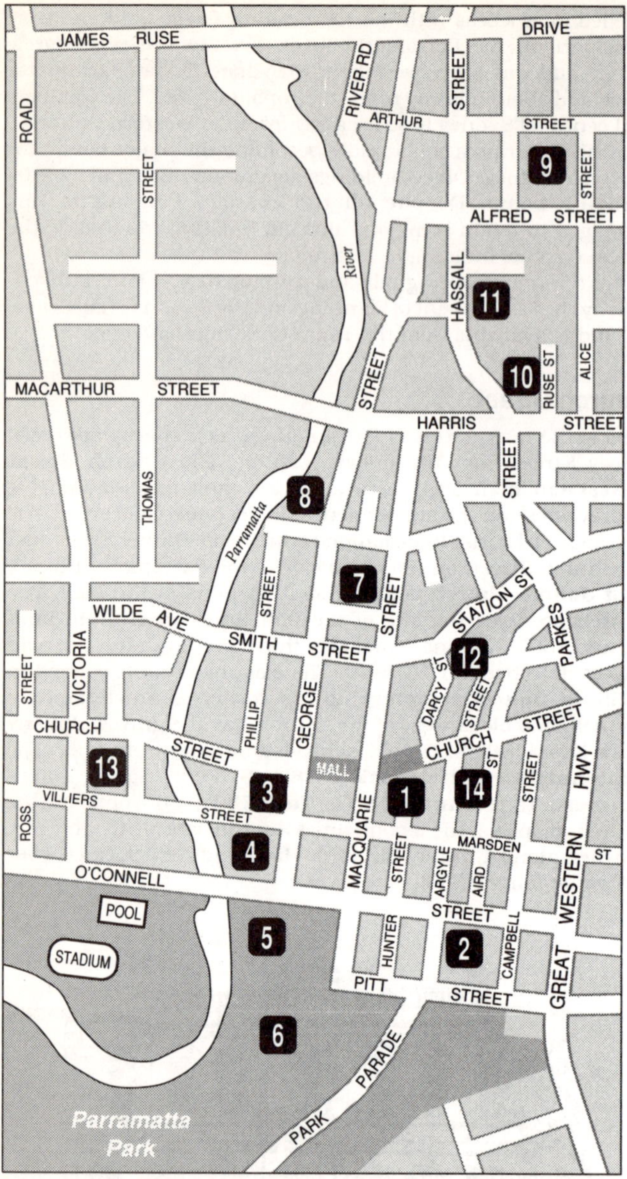

Parramatta

1   St John's Church
2   St John's Cemetery
3   Old Courthouse Tower
4   Brislington
5   Tudor Gatehouse
6   Old Government House
7   Perth House
8   Harrisford House
9   Elizabeth Farm

10   Experiment Farm Cottage
11   Hambledon Cottage
12   Parramatta Railway Station
13   Tourist Office
14   Westfield Shopping Centre

0   300 m

Tudor Gatehouse

Turning west along George Street, towards Government House, the first block contains **Woolpack Hotel**, and the old **Court house**. The Woolpack Hotel received its first liquor licence in 1798. The existing building was built in 1889 but it has retained a section of a building dating back from the 1820s. Although the interior is nothing special, it is not a bad place for a counter lunch.

On the corner of George and Marsden Streets is a fine Georgian colonial style building known as **Brislington** (1821). The original use is not clear, but may have been a hotel. It now houses a medical and nursing museum open on Sundays from 10 a.m. to 4 p.m. At the end of George Street is the **Tudor Gatehouse** (1885), built as the gatehouse for Parramatta Park, one gate for carriages, the other for pedestrians. For many years the building was used as a residence for park employees. The gatehouse is on the original site of the former entrance to Old Government House.

Parramatta Park

Parramatta Park is one of the oldest urban parks in Australia, dating from 1857. The winding Parramatta River and the many wonderful old trees make for a lovely park setting. A large public swimming pool complex on the north side of the Parramatta River has a pleasant picnic area. The Parramatta Stadium is a new and controversial addition to the park, located behind the swimming pool. Opponents of the stadium objected to potential traffic problems and damage to the character of the park. The Stadium is home for the Parramatta Eels rugby league team. Bicycle and boat rentals are available in the park.

Old Government House, beyond the Tudor Gatehouse in Parramatta Park, was the inland residence of colonial governors. However, in 1845 when the new Government House was built in the Domain, this house was no longer used. The building deteriorated badly until just after the turn of the century when major repairs were undertaken. Sections date from 1799 but it was extensively renovated in 1815 by Governor Macquarie. He succeeded in creating a gracious residence in the Palladian style with formal gardens. Behind the house is an interesting circular building. Originally the bathhouse (1823), it was converted to a pavilion in the 1880s. Water was pumped in lead pipes from the Parramatta River. The Old Government House is now a museum and has an extensive

Harrisford House

collection of early 19th-century furniture. It has a pleasant coffee shop. See Part 8 for hours and admission.

George Street, East of Church Street

In sharp contrast to the simple Georgian style buildings of the early 19th century is the **Roxy Cinema** (1930). The cinema is in the Spanish Mission style, designed by L.F. Herbert and E.D. Wilson. The elaborate facade has Spanish tile, wrought iron lamps and Spanish-style gates. The interior also maintains the Spanish theme.

On George Street, at Barrack, is **Perth House** (circa 1840). Another fine sandstone house, it is typical of buildings that graced George Street in the mid-19th century. One block further along at 182 George Street, is **Harrisford House** (1823-9). This lovely Georgian brick building was the original school rooms of King's School, an exclusive private boys' school with a reputation from its early beginnings for attracting the sons of pastoral families. The school is now located on the outskirts of Parramatta. Harrisford House is now a museum with period furniture, and is open Tuesday, Wednesday and Thursday, 10 a.m. to 3 p.m.

Queen's Wharf was just beyond the end of George Street, at Harris Street. The **HMAS Parramatta Memorial** now marks the spot where the wharf once stood. By the 1840s the river became so silted that it was no longer possible to run boats up to Parramatta.

Elizabeth Farm

Elizabeth Farm is located on Alice Street, at Alfred Street. It can be reached from George Street by turning south on Purchase Street (away from the river), then east on Hassall Street and south on Arthur to Alice Street. Elizabeth Farm was established by John and Elizabeth Macarthur in 1793. The Macarthurs conducted experiments in merino wool production. The colonial farmhouse was built in 1793 and parts of it remain, making it the oldest building in Australia. **Hambledon Cottage** (1824) on the corner of Hassall and Purchase Streets was built by the Macarthurs for their children's governess. The Elizabeth Farmhouse has been furnished in the 1820s to 1850s period and is open to the public. There is a kiosk serving Devonshire teas and snacks.

The **Experiment Farm Cottage** is on Ruse Street, a few

Elizabeth Farm

blocks south of the river on Harris Street. On the farm land surrounding this home, James Ruse was the first settler to harvest wheat after his planting in 1789. The house was built by Surgeon John Harris in the 1820s. The cellar of the house has early farm implements and the house has both local and English furniture from the early to mid-19th century.

Station Street

Just off Darcy street, at the railway station, is Station Street and here are two more very fine buildings. The **Lancer Barracks** (1820), built by Governor Macquarie, was the first military establishment in Australia. Part of the Lancer Barracks site is **Linden House** (1828). This house was originally located on Macquarie Street near Marsden Street, and moved to this site. It is used for a regimental museum, and is open Sundays 11 a.m. to 4 p.m.

Shopping, Restaurants, and Entertainment

Just off the south end of Church Street, under the railway bridge, is one of Sydney's largest indoor shopping malls with department stores, supermarkets, and a wide range of speciality shops. The David Jones department store is at the other end of Church Street, near the Parramatta River. There are several restaurants and cafes to choose from. Riverside Theatres, just off Church Street on the banks of the Parramatta River, is a complex of three theatres, including a cabaret theatre and restaurant.

PART V
The Harbour and the Beaches

On a fine sunny day, the sight of Sydney Harbour from one of the many headlands is breathtaking. Surely here is one of the most wonderful sights of any city in the world. At your first opportunity jump onto any one of the numerous ferries that ply the harbour and explore the hidden bays and beaches. And when the hot summer sun makes you long to cool off, what could be better than a dip in the surf of one of Sydney's numerous beaches? Exploring the harbour and beaches is one of the great treats awaiting you in Sydney.

16   Exploring the Harbour

Nothing distinguishes Sydney more than its wonderful harbour. Not so many years ago the harbour was full of magnificent sailing ships carrying wool and grain out of Australia and bringing goods in. Those romantic clippers are gone but you can nearly always see a big container ship slipping past the Opera House. Today the container ship traffic is considerably less compared with the days before the terminals at Botany Bay were built. However, the harbour is still full of ferries, water taxis, and pleasure craft of all kinds.

Because of its very special character, you can explore the harbour from the water and the land. There's a harbour shoreline totalling 244 kilometres including Middle Harbour, North Harbour and Sydney Harbour. To discover the harbour from the water, the action begins at Circular Quay. The public ferries are a good, leisurely, low cost way of touring the harbour. Travelling by ferry will allow you to get off and explore the quiet, village-like atmosphere of the north shore. If time is limited or you want a running commentary on the homes of the rich and famous in the eastern suburbs, then take a guided harbour tour. Any of the harbour trips may be made conveniently—but expensively—by water taxi. Ring 552 2266.

Finding the Best Views

To get your bearings, start with a trip to Circular Quay, the Bridge, or Mrs Macquarie's Chair. From any of these points, you will be able to see the main features of the harbour.

The area to the west of the bridge is the Parramatta River. It's narrower at this point but still large enough for big ships. There is a container depot at Drummoyne on the south, and oil storage tanks to the north. To the east the harbour is a series of bays, all the way to North and South Heads at the harbour entrance.

Mrs Macquarie's Chair at the northern tip of the Domain provides one of the finest views of the harbour. From this point the Opera House is framed by the Harbour Bridge. Kirribilli House and Admiralty House (the Sydney homes of the Prime Minister and the Governor-General respectively) as well as Neutral Bay are just across the water. Under the shade of the magnificent fig trees, the passing pageant of ferries, pleasure craft, and the occasional container ship will be enough to occupy you for hours. Access is from the Art Gallery (Explorer

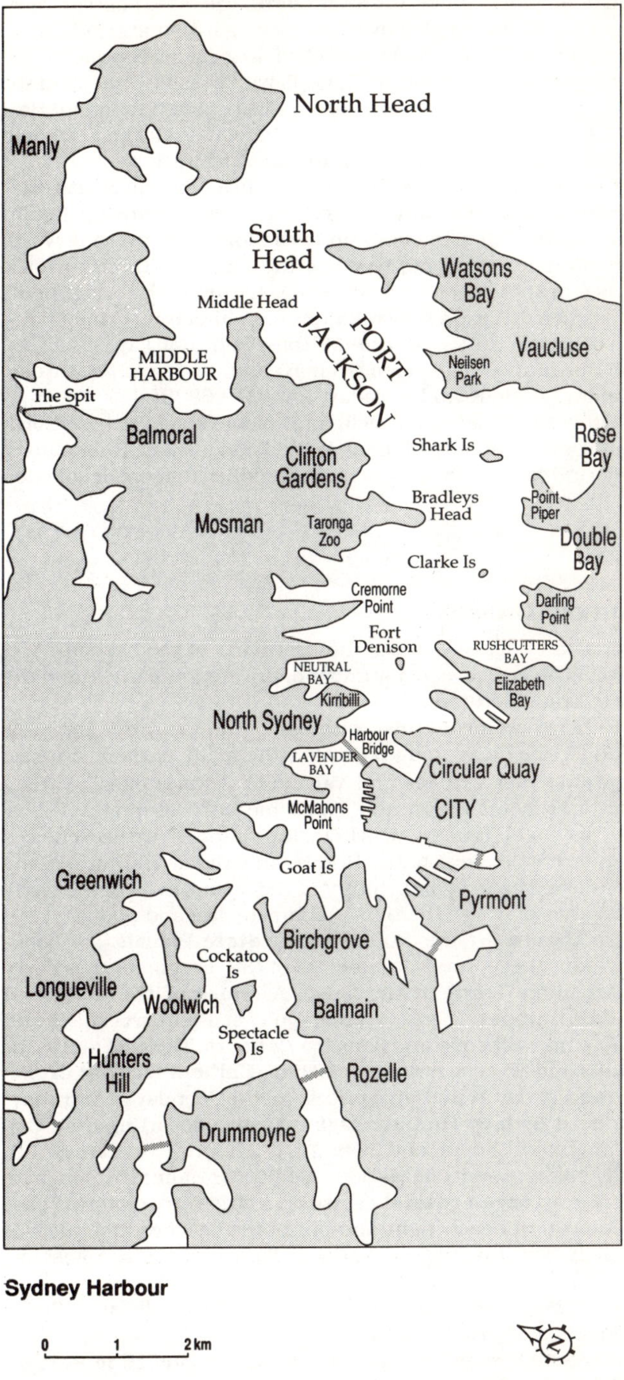

North Head

Manly

South Head

Watsons Bay

Middle Head

Vaucluse

MIDDLE HARBOUR

PORT JACKSON

Neilsen Park

The Spit

Balmoral

Rose Bay

Clifton Gardens

Shark Is

Bradleys Head

Point Piper

Mosman

Taronga Zoo

Double Bay

Clarke Is

Cremorne Point

Darling Point

Fort Denison

RUSHCUTTERS BAY

NEUTRAL BAY

Kirribilli

Elizabeth Bay

North Sydney

Harbour Bridge

LAVENDER BAY

Circular Quay

CITY

McMahons Point

Goat Is

Greenwich

Pyrmont

Birchgrove

Cockatoo Is

Longueville

Woolwich

Balmain

Spectacle Is

Hunters Hill

Rozelle

Drummoyne

Sydney Harbour

0   1   2 km

Bus, or free bus 666), or walk through the Domain from Macquarie Street or from the Opera House.

On the north shore, **Bradfield Park**, the most popular viewing point is now spoiled by earth-moving machines working on the harbour tunnel. Still untouched, just east of Bradfield Park is **Mary Booth Lookout** in Kirribilli with even finer views of the Opera House. The big ships seem close enough to touch. Best access is by Hegarty's Ferry at Circular Quay. Get off at Beulah Street and take the first left into Waruda Street.

For an aerial view of the harbour, head for the **South-east Pylon of the Harbour Bridge** (the pylon nearest Circular Quay). The 200-step climb up to the top of the pylon is worth the effort for sweeping views of the harbour. There's a little museum at the top with fascinating pictures of the bridge being constructed. Access is by walking only, either from Cumberland Street in the Rocks, or from Milsons Point Railway Station on the north shore. It's open from 10 a.m. to 5 p.m., Saturday, Sunday, Monday and Tuesday. Ring 218 6988.

Other good views and delightful picnic spots include North and South Heads, McMahons Point, Watsons Bay, Rushcutters Bay, and Neilson Park, which are described in more detail later.

Harbour Cruises

This is a quick way to see the best parts of the harbour. You have a choice between a number of private companies, and the State Transit cruises.

Either combine a cruise with one of the many bus tours, or head for the Quay and take one of the many harbour cruises. **Captain Cook Cruises** and **Vagabond Cruises** offer a variety of cruises, some with lunch, and some with morning coffee, as well as night cruises which include dinner. Cruise choices include east or west of the Bridge, and Middle Harbour. Many include a detailed commentary of the sights, even to a Beverly Hills style of where the rich and famous live and play.

Good value are the cruises run by **State Transit**. For bookings on these tours phone 27 4738 or go to the Ferry Information Centre at the Quay. A **cruise of the main and middle harbour**, leaves Circular Quay at 1 p.m. weekdays and 1.30 p.m. weekends and lasts 2 1/2 hours. The cost is $10 for adults and $5 concessions. A 10 a.m. **Harbour History Cruise** runs every day at the same prices. Nightly Monday to Saturday, there's a **Sydney Harbour Lights Cruise** departing from Circular Quay at 8 p.m. and lasting 1 1/2 hours. Costs are $7.50, and children and concessions $4.50. On summer Sundays, there is an **ocean coastal cruise** on a large ferry motoring past the northern beaches, in-and-out of the beaches and inlets of Broken Bay, Ku-ring-gai National Park and the Hawkesbury River. The round trip takes about 5 hours and leaves at 1.15 p.m. from Circular Quay. The cost is $30 for adults and $20 for students and pensioners.

A special Sydney treat is to follow the famous **18 footers** on a Saturday afternoon, during the racing season (November to March) These three-crew skiffs are reputably the fastest racing skiffs in the world, and it's easy to believe. A ferry leaves from the Quay and also from the Royal Yacht Squadron in Kirribilli at 1 p.m. The ferries have light refreshments and poker machines.

During the Festival of Sydney each January, a **ferrython** using the public ferries is held. The ferries are packed with race spectators and with luck you may book a ticket to follow the ferries around the course. Phone State Transit on 27 4738, or go to the Ferry Information Centre at the Quay.

Kirribilli, Neutral Bay and Mosman

The Neutral Bay ferry leaves Circular Quay for the 45 minutes round trip to Kirribilli and Neutral Bay. The first stop is **Kirribilli**. A pleasant walk from here will take you past Admiralty House (Governor-General's Sydney residence), and Kirribilli House (the Prime Minister's Sydney residence), and a 10-minute walk to the train station at Milsons Point. The ferry goes on to **Neutral Bay**, passing the Royal Yacht Squadron and **HMAS Penguin**, the submarine base. You will probably be able to see the black outline of at least one docked submarine. Buses meet the ferry at Hayes Street, Neutral Bay, if you wish to ride to the Neutral Bay shops on Military Road.

Before returning to Kirribilli and the Quay, the ferry stops at Kurraba Wharf. This is the starting point for a beautiful harbourside walk, if you have an hour or so to spare. From the top of the steps at Kurraba wharf, turn right if you want to stop at the Spains Lookout. Otherwise, turn left, take the first right (Shell Cove Road), first right again (Honda Road), and right again at Bogota Avenue. You are now looking at Shell Cove, one of the most beautiful and unspoiled of the many harbour coves. A walkway leads from Bogota Avenue along the shore of Shell Cove. The trees are thick and overhanging. A short but magnificent walk, with the most glorious views of the harbour and the Opera House, will bring you to Cremorne Wharf and a ferry back to the Quay.

Instead of rushing back to the city, climb the steps above the wharf to Robertsons Point. This is a favourite spot for rock

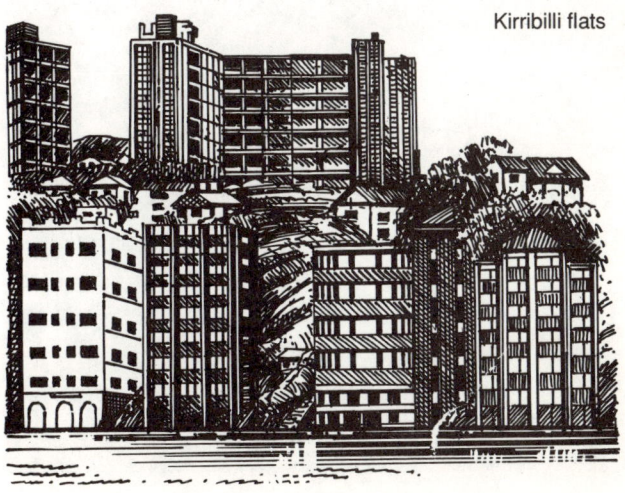

Kirribilli flats

fishermen. The small, white lighthouse makes a pretty picture. To your left you can see Mosman Bay and, further around, Taronga Zoo and the Sydney Harbour National Park at Bradleys Head. Follow the park walkway around **Mosman Bay**. On the one side are fine old houses with attractive gardens, and on the other the Moreton Bay figs lining the bay. Watch out for flights of squabbling rainbow lorikeets, and the occasional glimpse of crimson rosellas. Catch the Mosman ferry at the head of the bay. It's an easy walk all the way round to Bradleys Head if you find the temptation too strong to stop, although a street map would save some false turns.

Ferries to **Mosman** and **Taronga Zoo** take about 45 minutes for the return trip. Buses meet the ferry at Mosman Wharf if you would like to visit this picturesque habourside suburb. From the Taronga Park Wharf you can enter the Zoo, or take a bus up the hill to Mosman again. From the Taronga Wharf with a good sense of direction and a little perseverance, you can walk through the National Park at Bradleys Head, on to Chowder Head and Clifton Gardens. Part of the magic of these wonderful walks is being able to see the distant city across the harbour, yet seemingly being in the depths of the bush.

Lavender Bay and McMahons Point

For romantics who want to feel the splash of the sea, take Hegarty's Ferry from Circular Quay. Routes vary according to the time of day, but in the off peak, the ferry normally goes first to **Kirribilli** (Beulah and Jeffrey Street Wharfs), and then to Lavender Bay and McMahons Point. Get off at Beulah Street for a stroll around Kirribilli, or to stop at the Mary Booth Lookout. If you want to walk back along the Bridge, then Jeffrey Street is the closest wharf to the steps leading to the bridge (the steps are opposite the shops, just before Milsons Point Station).

Lavender Bay still has the feeling of time having passed it

Lavender Bay (*Ross Brownscombe*)

by. There are fishing boats moored here, and despite the increasing rate of development, Lavender Bay with the railway siding has a strange feel of being stuck in the early 1900s. At **McMahons Point** (Blues Point is just a little further around, but on the same headland), you can lie on the grass and contemplate the Opera House framed by the Bridge, or climb up to Blues Point Reserve and view Balmain and Goat Island in the distance. A short stroll up the hill will take you through an interesting hodgepodge of mostly 19th century houses and cottages. Blues Point has a good deli if you want to picnic by the harbour, or the excellent Blues Point Cafe for great meals, coffee, and cakes. A longer stroll, past pretty cottages, and narrow, twisting laneways, will take you to the **North Sydney** commercial centre. Hop on the train at North Sydney for a quick trip back into the city.

Hunters Hill, Balmain, and Birkenhead

To explore **Hunters Hill** take the ferry to Woolwich Wharf, and take Bus No. 234 into the village. It's hard to believe this serene neighbourhood survives in the middle of such a large city. A great place for a quiet stroll back to the 19th century. There are some wonderful old sandstone buildings, beautifully preserved, as well as the old town hall. Check under Hunters Hill later in this section for details. The same ferry also goes to Balmain, so you can easily stop off on the return trip.

The **Balmain** ferry also goes to Greenwich and Hunters Hill. Get off at either Long Nose Point or Darling Street Wharves. Buses 442, 445, and 434 leave from the Darling Street Wharf. Explore the village atmosphere, take in a meal or enjoy one of its pubs. Note, outside of peak hours it is wise to tell the deck hand if you want to get off at Long Nose Point.

For something a little bit different, take a bus to the **Birkenhead shopping centre** (Nos 500, 501, 502, or 508 from Circular Quay). The shopping centre includes all sorts of food shops, and shops selling souvenirs and bric-a-brac. Fast food is a big attraction here as are the fruit and vegetable markets. Other attractions include the Lego Centre, open every day from 9.30 a.m. to 5 p.m., ring 819 6888.

Meadowbank

An enjoyable way to explore the **Parramatta River** west of the Harbour Bridge is to take the ferry from Meadowbank. The ferry leaves Circular Quay on weekdays, every two hours. For a quicker return to the city, walk the 10 minutes to Meadowbank Station and take the train back to the city.

Manly and Watsons Bay

The trip to Manly is one of the great Sydney traditions. The ferry trip takes about 30 minutes and is the easiest way to see the length of the harbour. Make sure to watch out for the harbour heads in the distance to catch a glimpse of the Pacific. Look out on the north side for the beautiful Middle Harbour. The headlands are largely free of development and unspoiled so give some idea of what Cook, and the First Fleet must have seen 200 years before. On a Saturday or Sunday the harbour is alive with pleasure craft, and on a Saturday afternoon, during the season, with luck you will catch the 18-footers racing. Ferries leave every 30 minutes. The trip can be done in less than 20 minutes by hydrofoil.

At weekends ferries go hourly to **Watsons Bay** from the Quay, Taronga Zoo and Rose Bay. From Watsons Bay, enjoy the walk along the beach to Camp Cove, or swim from the nudist beach at Lady Bay, or climb up onto the cliffs for a dramatic view of the ocean at South Head.

The Harbour Islands

There are eight islands in the harbour. Several of them have been popular picnic spots for years. In the 1920s row boats could be hired for two shillings an hour from Rushcutters Bay for picnickers going to Shark and Clark Islands. Steamers also stopped at these islands on weekends. These days tours are run to Fort Denison and Goat Island but as for the others you'll need to have your own means of water transport or hire a water taxi. State Transit runs a five island cruise which leaves the Quay at 10 a.m. daily, to see, but not to visit, Fort Denison, Shark Island, Clark Island, and west of the Bridge, Goat, Cockatoo, and Spectacle Islands.

Fort Denison, also known as **Pinchgut**, was a rocky peak in the harbour just off Farm Cove. Convicts, isolated on the island, built the fort. One convict was executed here for murder. Joseph Holt, a convict who arrived in Sydney in 1880, described his first impressions of Sydney:

> Just at day-light, we entered Sydney Heads, and fired a gun for a pilot but none appeared; we sailed by Pinchgut Island. The first remarkable object I saw was the skeleton of a man, Morgan by name, on a gibbet...

The skeleton was left hanging there for three years. Fort Denison was used to help defend Sydney against a feared invasion from the Russians during the Crimean War (1854-56). Pinchgut Island was levelled for the base of the fort. At Fort Denison you can climb to the top of the Martello Tower and get splendid views of the harbour. There are still cannons here as well as the powder room and the tide rooms. The Maritime Services Board runs tours to Fort Denison from Tuesdays to Sundays, departing at 10.15 a.m., 12.15 p.m. and 2 p.m.. Tours last 90 minutes. Bookings are essential and can be made by calling 272 733. The cost is $7 for adults and $4.50 for children and pensioners.

Goat Island is west of the Harbour Bridge, near the tip of Balmain. Early settlers used the island for grazing goats. Now it's a ship repair yard but there are still many remnants of the early colonial days. An old quarry, begun in 1831, provided quarry stone for government buildings. As many as 200 convicts worked here. A powder magazine, constructed in the 1830s, is located on the floor of the quarry. Thirty-three metres long, the walls are a massive 3 metres thick. Still standing beside the magazine is the cooperage which made the barrels for storing gun powder. The barracks where British ordinance soldiers lived have been turned into the Maritime Services Board harbour museum tracing over 250 years of maritime history.

The Maritime Services Board runs tours to Goat Island from Wednesday to Sunday at 10.45 a.m. and 12.45 p.m., lasting about 90 minutes. Bookings are essential; phone 272 733. The cost is $7 for adults and $4.50 for children and pensioners.

East of the Bridge are **Shark Island** and **Clark Island**. As no ferries stop here, the most likely transport is water taxi or private boat.

Rodd Island is west of the Bridge, in Iron Cove near Drummoyne. A turn-of-the-century dance hall and some summer cottages are reminders of livelier days on Rodd Island. Now, alas, the entertainment is limited to group picnics, and the only facilities are picnic tables and toilets. Rodd Island is run by the Sydney National Park, and is open on weekends from November to April. Make advance bookings by calling 337 5355. No ferries go to the island, so travel by water taxi. Sailboard rentals are available at Rodd Park on Henley Marine Drive on the bay a few hundred metres from the island at Rodd Park. The easiest way here is a 494 bus from Burwood Station, along First Avenue to Rodd Park.

17   Rushcutters Bay to Watsons Bay

The 324 and 325 buses marked "Watsons Bay" leave from the Quay via Pitt/Castlereagh Streets and Kings Cross. The buses also stop at Edgecliff Station on the Eastern Suburbs line. Both go close to Rushcutters Bay, Double Bay and Rose Bay. The 345 then goes on past Neilson Park.

The first bay after Kings Cross is **Rushcutters Bay**. In the earliest years of the colony, the rushes growing here were used to make ladies' hats. The lovely bay is framed mostly by apartment buildings. On a Saturday afternoon during the season, many of the 18-foot racing skiffs are launched from here. On the east side of the bay, is the Cruising Yacht Club of Australia. The racing fleet of the Sydney to Hobart ocean classic sets sail from here every Boxing Day.

In the park on the east side of Rushcutters Bay on New Beach Road, Duncan MacLennan's Boomerang School gives boomerang throwing lessons every Sunday from 10 a.m. to 12 noon. Their shop is at 200 William Street. For a look at the neighbouring bay to the west, **Elizabeth Bay**, there's a steep path from the western end of the park over Macleay Point. **Elizabeth Bay House** is on the slopes of Elizabeth Bay on Onslow Avenue. See Part 8 for a description and opening hours.

Travelling east from Rushcutters Bay you are now in the

Elizabeth Bay House

heart of the "eastern suburbs". This area probably has the highest concentration of expensive residential real estate anywhere in Australia. Famous harbourside suburbs include Point Piper, Darling Point, Double Bay and Vaucluse.

Double Bay is known for its exclusive shops, and its cafes and restaurants. If you're after international fashion labels this is the place to come. You'll also find everything from hand-embroidered lingerie at White Ivy on New South Head Road to Australian jigsaw puzzles crafted in silver at Shiel Abbey's on Cooper Street.

Double Bay is seen best by just wandering around. Outside of the commercial area the surrounding neighbourhoods are worth a look. Bay Street will take you down to the harbour. At the bay shores there is a lovely little park, ideal for a picnic if you've stocked up on bagels at one of Double Bay's delis. Savour the atmosphere of Double Bay over a coffee and cake at George's or "21" or "Cosmos". The Cosmopolitan is open to 2 a.m. every day and their full menu is served at all hours. There are several restaurants here and this is the place if you're looking for kosher dining.

Next along the harbour is **Rose Bay**. The Woollahra Sailing Club and the Royal Motor Yacht Club operate from here. Not so long ago the great Sunderland flying boats left from Rose Bay to Lord Howe Island. When an airstrip was built on Lord Howe Island in 1975 this service was stopped. There is still a small base here for flying boats.

The 325 bus will drop you at Greycliffe Street for **Neilson Park**. You can swim from the little beach here (shark-netted in the summer), or picnic in the lovely grounds. The views of the harbour from here, back to the city, are stunning. The view across the harbour is mostly of the bush reserves in Mosman.

Vaucluse and **Parsley Bays** are intimate little bays with small beaches. **Vaucluse House** is in Vaucluse Park (bus 325 stops at the front gate). Built for the explorer, Charles Wentworth in 1803, the gothic style is replete wih turrets. The splendid setting has been preserved to the water's edge. The house is furnished in period furniture and is open to the public. "Aussie Tea" is available in the tea rooms. They serve such delights as herb damper, pickles, chutney and King Island cheeses. Devonshire teas are also available. See Part 8 for admission and hours.

Lastly along the harbour is **Watsons Bay**. Within a distance of 200 metres you can choose between the windswept cliffs with the crashing waves of the open ocean and the quiet waters of the harbour, with the shade of giant fig trees.

The view from the cliffs across the street from the park is spectacular. Known as the **Gap**, just down from the harbour's South Head, the waves break on 50-metre cliffs. The Gap is Sydney's most popular scenic suicide spot where a couple of dozen people every year make it their final leap. Take the path along the top of the rocks. The anchor is a reminder of the ferocity of the sea. The immigrant ship the **Dunbar**, sailing to Sydney in 1857, ventured too close to the rocks and all but one of its 122 passengers drowned. The wreck of the **Dunbar** is now a favourite of adventurous scuba divers.

To savour the full atmosphere of this quiet corner of the city, try **Doyle's Seafood Restaurant**. One of Sydney's most famous restaurants, it is renowned for its fish and harbour views.

Along the bay front a few old weatherboard cottages have miraculously survived, despite the fact that land values must

Doyle's at Watsons Bay

be over $1 million. The neighbourhood of Watsons Bay makes for a pleasant walk, with its narrow streets and quaint huddle of houses. Cliff Street ends at **Camp Cove**, where there is a quiet beach and a park reserve. If you follow the path up the metal steps on the north side of Camp Cove you'll come to **Lady Bay**, where nude bathing is permitted.

The tip of **South Head** is taken up by the naval base, **HMAS Watson**. Visitors are allowed to visit the naval chapel with its lovely stained glass mosaic.

Along Old South Head Road, south of the Gap, is a handsome lighthouse designed by Greenway in 1818. The views from this location are wonderful, both back over the city and out to sea.

18   Lower North Shore: Hunters Hill to Mosman

The character of Sydney's north shore is quite different from the rest of Sydney. Yet, it has such variety, you will be busy for days exploring its hidden corners. The overall appearance of the north shore, save for the high-rise buildings of North Sydney and Kirribilli, is that of leafy, quiet suburbs.

On this part of our trip we start at the historic and tranquil streets of Hunters Hill, then to North Sydney, Australia's fourth largest office centre, and on to Taronga Zoo in its superb peninsula position. All these places can be reached by ferry.

Hunters Hill

Take the ferry from Circular Quay to the Valentia Street Wharf (also known as the Woolwich Wharf) in Hunters Hill. Bus 234 from the North & Western Bus Lines, should meet the ferry. Take this to Alexandra Street. In peak hours a commuter ferry runs to the more convenient Alexandra Street Wharf. The centre of Hunters Hill is only a short walk along Alexandra Street. There are no ferries on Sundays. A 506 bus from the Quay will take you all the way to Hunters Hill.

Hunters Hill is off the beaten track, perfect for a quiet stroll. Leafy streets are lined by fine sandstone 19th century homes. The entire municipality of Hunters Hill is classified by the National Trust. The first houses were built in the 1850s. Didier and Jules Joubert from Bordeaux, France were the first developers and built stone houses with a distinctly French quality.

Hunters Hill was not always a suburb strictly for the affluent. Many of the original stone cottages were only two rooms. Residents now are busy adding extra rooms, bathrooms, skylights and sunrooms, but thankfully most have retained the original facades. Especially interesting streets are Alexandra, Stanley, Ady, Madeline, Mount, and Ferry Streets. For a detailed walking tour you can pick up a pamphlet at Hunters Hill Town Hall during business hours.

Alexandra Street is the main street of Hunters Hill and despite all the changes retains much of the character it had in the late 19th century. Stone cottages line the north side of the street and stone villas stand on the south side. The stone kerbing and walls date from the 1860s. Here also are interesting shops and the oldest pub. The Congregational Church and Hunters Hill School on the corner of Stanley Street, were both built in the 1870s. Further down is the stately Town Hall built in 1866, and across the street is the elegant "Merimbah" home. "**Vienna Cottage**" at 38 Alexandra Street is a workman's cottage restored by the National Trust. Inside is a display on the history of Hunters Hill. Open hours are 2 p.m. to 4 p.m. on Saturdays, and 11 a.m. to 4 p.m. on Sundays, or by appointment: ring 816 2255. Across from Vienna Cottage is an attractive cluster of shops and cafes in 19th-century buildings. "Cuneo" dates from 1883, and the Garibaldi Inn was the first inn in the area, built in 1861.

Back towards the school is **Stanley Street**. "Eulbertie" (1857), originally a doctor's residence is now used as part of the Hunters Hill school. "Loombah" at 3 Stanley Street (1879) is a gracious Georgian-style mansion, and "Lyndcote" at 7 Stanley Road (1858), is an elegant home fronting the Parramatta River.

Kirribilli, Lavender Bay and McMahons Point

Milsons Point Station is a starting point for most visits. Just one stop on the train from Wynyard, the transformation from the noise and fumes of the city is dramatic. Alternatively, a leisurely ferry ride from the Quay helps to set the tone. One fun way to get to this part of the city is to walk across the Harbour Bridge. Stairs lead up from the Argyle Steps in the Rocks and at the north end the stairs are located near the Milsons Point railway station. The walk takes about 15 to 20 minutes. It's noisy with all the bridge traffic but the views make it well worth while.

For **Kirribilli** you have a choice between the State Transit ferry to Kirribilli Wharf, or Hegarty's ferry to Jeffrey Street or Beulah Street wharves. Kirribilli is an enchanting mixture of the old and new. High rise buildings dominate the skyline, but many of the old terraces and pre-war flats survive. It's a sought after suburb in which to both buy and rent because of its superb location. The **Prime Minister's residence**, Kirribilli House, as well as the **Governor-General's** at Admiralty House, are found here, commanding the south-east corner of Kirribilli. They are only open on special occasions.

One of the attractions of Kirribilli is the Kirribilli Fish and Chips shop, on Fitzroy Street. Take-aways only, the nearest park is back towards the harbour where you get a lovely view of the skyline, Bridge and Opera House. There are also a few coffee shops, restaurants and pubs in the centre. A walk down McDougall Street will take you to Milson Park on Careening Cove, another pleasant park. There's the **Ensemble Theatre** here, right at dockside. Some of the 18-foot racing skiffs also set sail from here on race days. At the end of Elamang Avenue on Peel Street, is the Royal Yacht Squadron, tucked in behind a forest of masts. The restaurant and bar are pleasant and overseas visitors don't need to be accompanied by a member.

On the west side of the rail station is **Milsons Point**. Now a burgeoning office centre as an off-shoot of North Sydney, it has lost most of the charm that it once had. The big attraction here is the North Sydney Olympic Pool; a lovely 50 metre pool with children's wading pools. Part of Australian swimming history, many great Australian swimmers of the 1950s trained here. In the winter months a plastic bubble is installed over the pool and it remains open all year round. There's a large grandstand that gives you great harbour views while sunning yourself. And when in the pool the rattle of the trains on the Sydney Harbour Bridge adds to the ambience. There's a good snack bar where you can get breakfast. The pool is a popular place with the exercising office workers. Open 6 a.m. to 8 p.m. Monday to Friday, and 7 a.m. to 6 p.m. Saturday and Sunday.

The most famous attraction of Milsons Point until recently was **Luna Park**. The fun park is now closed, but the lovable Luna Park entrance clown face has been classified for protection by the Heritage Council.

Lavender Bay and McMahons Point are the next two harbourside suburbs west of Milsons Point. From the Ex-servicemen's club on Cliff Street, steps lead down to Clark Park, for great views of the Bridge, and then under the railway bridge to the water's edge, and the Lavender Bay wharf. Wind your way up to McMahons Point from here. Hegarty's ferries stop at both Lavender Bay wharf and at the end of the point.

McMahons Point still retains much of its character, despite some new apartment blocks. The prime position on the Point is occupied by Blues Point Tower, designed by Harry Seidler and completed in 1961. Architect Harry Seidler and friends in the late 1950s proposed to flatten all the houses on the Point to make way for high rises. Thankfully this proposal didn't progress past the drawing board.

In McMahons Point "village" there's a small collection of shops and restaurants. The "Potiniere" has delicious french food, and other attractions include a wine bar, The Grape Escape. The popular "Blue's Point Cafe" can't be beaten for its tasty food at reasonable prices and they have a take-away service too.

A walk through the charming streets of McMahons Point almost always leads to a lovely harbourside park, with some of the best views of the harbour and the city. The park at the end of the Blues Point Road is a great place for a picnic. If you're seeking shade climb up to the park right at the point, beyond the large apartment building. At the end of Henry Lawson Avenue, by the ferry stop, the "Harbourside Hotel" has self-contained apartments available as tourist accommodation. "Sails" Restaurant is right on the wharf.

North Sydney

Take the train on the North Shore line at North Sydney station and head up the hill to Miller Street, or by bus from behind Wynyard Station.

In the last fifteen years, **North Sydney** has been transformed into Sydney's second business district. You'll be hard-pressed to find a commercial building dating back much before the 1950s. The office boom started in the 1970s and although activity is slowed, there always seems to be a crane on the horizon. More than 40,000 people work in the North Sydney centre, which is especially popular with advertising firms, computer, and consulting firms. Not surprisingly, with such a concentration of high-living yuppies, there's no shortage of restaurants and bars.

The main shopping mall is tucked away behind Miller Street, enter from Berry Street, or Mount Street.

One of the few buildings with any style in North Sydney centre is the mock Spanish Tower Square. Mostly fast food, restaurants and cafes, the outside tables are crowded at lunch times. When the clock in the tower strikes the hour, a bull fighter and bull appear from the clock. In amongst the offices is the Walker Street Cinema, a good second-run film cinema. On Napier Street, tucked in behind the Pacific Highway on the west side is little **Don Bank Museum**, see Part 8 for details. And across from the museum is Callaway's, a great spot for a drink or a meal.

Neutral Bay, Cremorne, and Mosman

The most enjoyable way to travel is by ferry to Neutral Bay, Cremorne or Mosman, and take the connecting buses to the heart of the commercial areas. If you are in a hurry, buses 246, 247 and 249 from behind Wynyard are a quicker alternative out of rush hours.

The attractions of these leafy suburbs are the wonderful harbourside walks, the shopping strip on Military Road and in Mosman Village, and its elegant residential areas. Taronga Zoo, described in the next section, is in Mosman on the harbour shores.

Military Road is a long shopping strip starting in Neutral Bay, passing through Cremorne, and terminating in Mosman Village at the start Bradleys Head Road. The Military Road shopping strip is a sprawl of shops along a very busy road. Gradual piecemeal development has added a number of small shopping plazas and small office blocks.

The Mosman end of Military Road is much more pleasant as it has more of a village feel. The wealthy suburbs around this area have created the demand for a great variety of shops, ranging from very exclusive clothing shops to shops selling teddy bears only. The shopping area is dotted with cafes and restaurants, pastry shops and delicatessens. There are some popular bars too, including "The Oaks", at Military and Ben Boyd Roads, Neutral Bay, and the "Metropole" at 287 Military Road, Cremorne.

Mosman is situated on a headland that protrudes into the

harbour. On the north side is Middle Harbour, which is smaller but even more dramatically beautiful than the main harbour. With a small boat you might spend all summer exploring the golden beaches and tiny coves. Enchanting and evocative names such as Chinamans Beach, Quakers Hat Bay, and Powder Hulk Bay, are poignant reminders of a rich history. If you have a chance to explore Middle Harbour by land or sea, you will see Sydney at its most spectacular.

Lovely **Balmoral Beach** is one of the local beaches for the north shore. It's a wonderful, long beach with lots of activity, but it's much quieter than Bondi. The beach front has just a few village shops, including take-aways and restaurants. You can swim in the netted baths area, or anywhere along the beach, rest under the shade of a fig tree, dine at the water's edge or rent a sailboat or windsurfer. On Sundays in summer there are jazz concerts in the rotunda. The water here is quite still, so ideal for sailing and swimming. The promenade along the beachfront is a pleasant place for a stroll.

One extraordinary thing about Balmoral's past is that in 1924 a large amphitheatre, true to the "Grecian Doric style", was built right at the water's edge on the north end of the beach. It could seat 3000 people and the stage rose 70 feet above the water. A religious group called Star in the East, sponsored this great work as a vehicle to spread theosophical beliefs. The classical theatre and public lectures could not draw the crowds and it soon stood empty. At one point it was even used for mini-golf. Demolished in 1951, a block of home units called Stancliff now stands on the site.

To reach Balmoral Beach by bus, take the Taronga Zoo ferry to Athol Wharf, and then bus 238 to the beach. You may also walk from Mosman, or take the 257 bus that leaves Chatswood Station, travelling to Balmoral via Mosman. If you're driving on a fine summer weekend, best make an early start as parking is a problem.

The bays and suburbs between Balmoral and the Spit Bridge make an enjoyable tour if you are driving. You might also like to take the trip to **Spit Bridge**, from where you can have a good view of Middle Harbour, and watch the boats. The 8-kilometre Manly Scenic Walkway starts from the Spit Bridge, if you have the time, this is strongly recommended. Check the Manly section for details. There are few shops at the Bridge, so best to bring what you need. Any Manly bus will take you to the Bridge, try the 144 from Chatswood or St Leonard's Stations, or the 182 and 184 from Wynyard, the Narrabeen and Mona Vale buses.

Taronga Zoo and Ashton Park

The zoo ferry leaves Circular Quay every half hour, and takes only 12 minutes, making it the easiest and most relaxing way to get to the zoo. Bus 247 leaves from Wynyard every 30 minutes during the week, but does not run on Sundays.

On arriving at the zoo wharf you can enter at the lower entrance, take the bus to the top entrance, or the "Aerial Safari" cablecar at a cost of $2.50 for adults and $1 for children. The zoo is open every day from 9 a.m. to 5 p.m. and admission is $11 for adults and $5 for children, students, and pensioners. Phone 969 2777.

Taronga Zoo is probably one of the most pleasant zoos in the

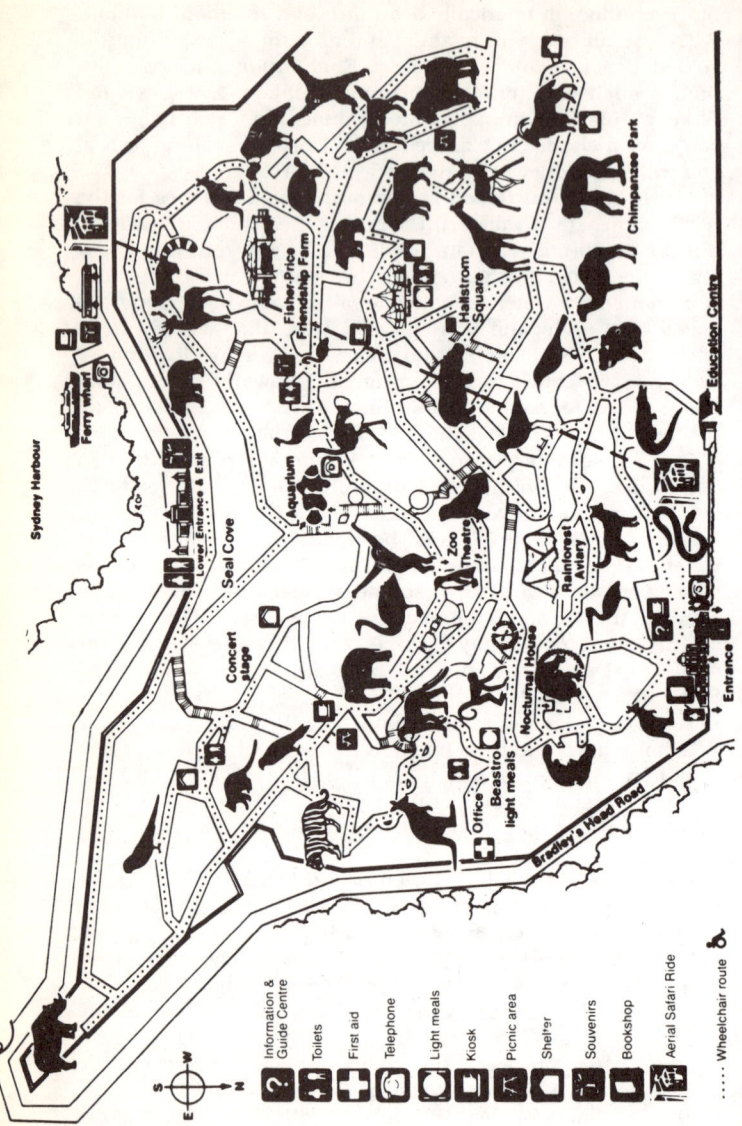

Taronga Zoo

world. Many other modern zoos have shed their "cement and bars" look for that "open and leafy" look. But what sets Taronga apart from all the others is its spectacular location, with an entire harbour peninsula devoted to the zoo grounds. The grounds themselves are a delight, with plenty of shade trees, flowers, and harbour views.

A big attraction at the zoo is the Australian animal exhibits, the largest in Australia. Joining one of the Australian Walkabout tours given by volunteer guides is a great way to see the Australian exhibits. The tours are informative and the groups are small enough to allow time to ask questions along the way. Many overseas visitors make a bee-line to the koala enclosure. A viewing platform winds around the koalas in their gum trees. To see these sleepy creatures actually stirring, be there at 3 p.m. feeding time. To see the extraordinary platypus in action, the best viewing time is 2 p.m. As many Australian mammals are nocturnal, they must be presented in nocturnal houses if the visitor is to see much activity. The nocturnal house, with its simulated moonlight allows you to watch what they would be doing after nightfall.

Australian Fur Seals can be found at Seal Cove and free and very entertaining seal shows are held at 1.15 p.m. and 3.15 p.m. daily, with an extra show on Sundays at 11.15 a.m. A special attraction is the large walk-in aviary giving you the feeling of being in the Australian bush but it's a lot easier to spot parrots, kookaburras, and bower birds in here.

All the traditional zoo attractions are also here. The monkeys and penguins are popular, and the large aquarium at the lower entrance has a colourful display of creatures from The Great Barrier Reef.

Ashton Park occupies the peninsula south and east of the zoo and is part of the Sydney Harbour National Park. Reach it by road down Bradleys Head Road or walking uphill from the Taronga Zoo ferry. The Park is rich in military history. Governor Gipps chose this strategic point to build a gun pit in the early 1840s. Fearful of attack by the Russian Fleet in 1871, a battery was added to the military fortifications. The Russians were never sighted in the harbour, but in memory of other naval actions, the mast of the **HMAS Sydney** now stands prominently in Ashton Park commemorating the sinking of the German raider Emden in 1914.

Hard to believe now, a rowdy dance hall and pub once featured on the harbour shore, and plans for the mining of shallow coal deposits in the park were cancelled only after public outcry. This beautiful park remains a rare and almost pristine example of vegetation of Hawkesbury sandstone within the urban area. Floral conservation techniques developed in Ashton Park by two local women, the Bradleys, have been applied widely in other sensitive areas across Australia. The bottlebrush, she-oaks, yellow pea flowers, irises and native orchids now thrive in this tiny enclave in the midst of the city, thanks to many years of devoted work.

Enjoy over 4 kilometres of easy walking trails, a lovely picnic area and sandy beaches. If you are lucky enough to be in Sydney on December 26, Ashton Park is a favourite viewing spot for the Sydney to Hobart Race. Less majestic, but just as exciting, the 18-foot skiffs sail close to the point on Saturday afternoons during the racing season.

19   Manly

One of Sydney's finest pleasures is the ferry ride to Manly. The 30 minute journey is packed with the sights of the harbour. If you cannot wait so long, take the hydrofoil, for a 15 minute ride. Far slower is the bus from Wynyard station. Buses also leave from the North Shore, take Bus 144 from St Leonards Station.

As the story goes, Manly gets its unusual name from Governor Phillip who thought the Aborigines here had a particularly "manly" air about them. Despite their fine qualities, they were not left to enjoy the views for much longer. Manly was soon to become a popular retreat from the grime of the city. Even in the mid-19th century Manly was a famous seaside resort. Back then swimming in the sea in daylight hours was forbidden, and not until the turn of the century did it become acceptable. Most of the other Sydney seaside resorts are pale shadows of their former splendour, yet Manly is thriving and still a popular destination with country people for their annual holiday. Manly still has much to offer as a holiday destination. With its ocean and harbour beaches, Manly is a great fresh air and sea breeze alternative to the inner city. Accommodation comes in all shapes and sizes, together with night life, fast foods and restaurants galore.

Manly's greatest attraction is its ocean beach, over 1 1/2 kilometres long. In fact, only rocky headlands jutting into the sea interrupt the chain of beaches which stretch all the way to Palm Beach at the end of the Warringah Peninsula. All you need for enjoying the beach—changing sheds, showers and surf board rentals—are in easy reach. Manly has a long surfing heritage. The sport was introduced to Australia here in 1915 by a visiting South Pacific islander. Manly stages the Coca Cola surf competition every year, usually in April; phone the tourist information centre for more information on 977 1088.

The **Tourist Information Centre** at the beach, opposite the Corso, is open every day from 10 a.m. to 4 p.m.. Information on accommodation, restaurants, transport, and attractions is available here. If you want to explore Manly in detail, there is an excellent little publication of the New South Wales Department of Planning and the Manly Municipal Council, entitled **Manly Walk**, available at 175 Liverpool Street, Sydney. A convenient way to visit the sites of Manly and North Head is the Manly Explorer sightseeing bus. Get on and off as you please for the price of one daily ticket. The Explorer leaves the ferry wharf every hour on weekdays and half hourly on Saturday and Sunday afternoons for only $2.50 for adults, and $1.50 children and pensioners. Phone 938 4677 for bus times, or ask at the tourist information centre.

The pedestrian mall, **the Corso**, has a selection of shops, pubs, restaurants, and take-aways. There are also some good seafood restaurants on the Esplanade. Star billing on the Corso currently belongs to a skateboarding bull terrier. He hangs out in front of the "New Brighton Hotel". Neither skateboarding nor dogs are allowed on the Corso so he posed a vexing problem for local authorities. The inspired solution to this bureaucratic teaser—a special busking licence! Don't be too disappointed if he's nowhere around, this dog has his sights on Hollywood.

Manly's great attraction is the beautiful beach and crashing

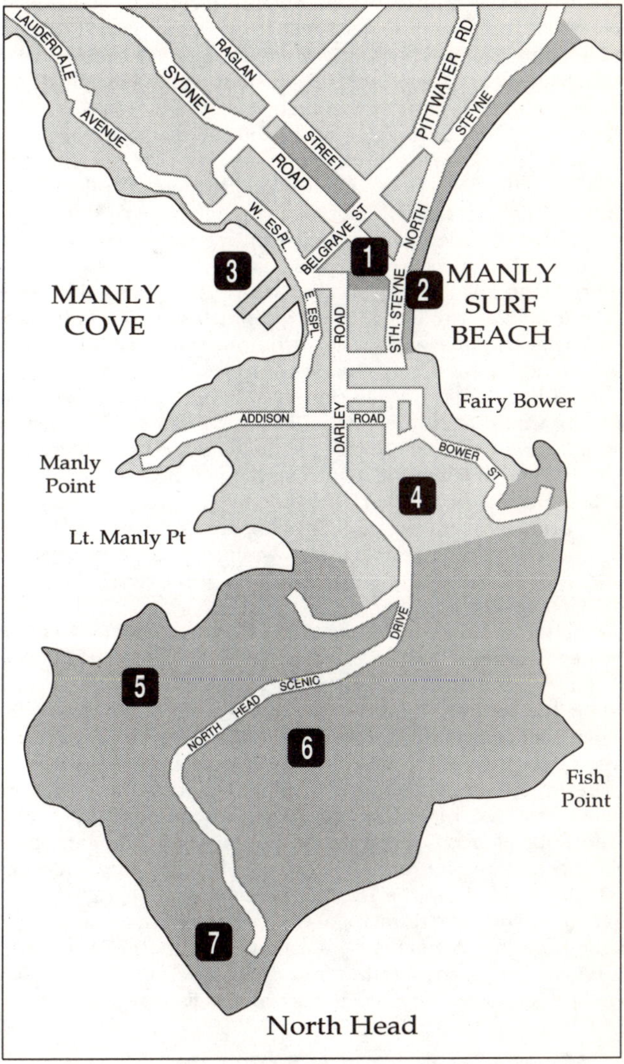

MANLY COVE

MANLY SURF BEACH

3

1

2

Fairy Bower

Manly Point

Lt. Manly Pt

4

BOWER ST

ADDISON

DARLEY ROAD

ROAD

5

NORTH HEAD SCENIC

DRIVE

6

Fish Point

7

North Head

LAUDERDALE AVENUE

SYDNEY ROAD

RAGLAN STREET

BELGRAVE ST

W. ESPL.

E. ESPL.

STH. STEYNE

NORTH STEYNE

PITTWATER RD

1 The Corso
2 Tourist Information
3 Manly Ferry & Bus Terminal
4 St.Patricks College
5 Quarantine Station

6 Military Reserve
7 Park Hill Lookout,
   Sydney National Park

0   0.5   1 km

surf, but for a break from the sun, see the shark feeding at **Manly's Underwater World**. Reputedly Australia's largest collection of marine life, see also the gropers, fur seals, octopi and many tropical fish. As part of the aquarium complex, The **Manly Waterworks** has four giant water slides. It's just west from the ferry wharf, up the hill and a 5 minute walk. See Part 6 for cost and opening hours. The **Manly Museum and Art Gallery** near the Underwater World on the harbour side, has good art exhibits as well as interesting memorabilia, see Part 8.

The grand stone building up in the hill at the ocean is St Patrick's College, a Catholic seminary. It's not open to the public. A walk along the beach in this direction will take you to the pretty Shelley Beach. There are great views over Manly and the sea from Shelley Point which can be reached by following the path at the end of the carpark.

For a longer walk, the **Manly Scenic Walkway** follows the harbour all the way from Manly to Spit Bridge. The walk is 8 kilometres and takes about 3 to 4 hours. Many sections of the walk cut across bushland reserves and even a section of Sydney National Park. The route is complicated, so before you set off pick up a map from the National Parks and Wildlife office in Cadman Cottage in the Rocks, or the Manly tourist office. The walk is one of startling contrasts, quiet beaches of Forty Baskets Beach and Clontarf, the brooding grandeur of Dobroyd Head, and the picturesque lighthouse at Grotto Point. Buses 182 and 184 to Spit Bridge from Wynyard in the city, or start at Manly wharf. For a long distance ocean walk, see the Northern beaches section.

Don't leave Manly or Sydney before a trip to **North Head** in **Sydney Harbour National Park**. North Head, marking the entrance to the harbour, is a sheer cliff, wind-swept and treeless at the top, plunging to the blue waters of the ocean far below. The views of the harbour and distant city skyline are unforgettable. Inside the park is the old **Quarantine Station** at Spring Cove. First opened in 1828, immigrants and visitors suspected of having contagious diseases were confined here as recently as 1984. Officers from the station would fumigate ships and check for rats up until 1972. Most of the existing buildings on the site date from the 1880s. There are also Aboriginal rock engravings from the Kameragal Clan who lived in the area. One of the buildings houses an enormous collection of European and Asian engravings dating from the early 1800s. Inspection is by guided tours only, see Part 8 for details.

20   Northern beaches

Bus 190 from Wynyard Station goes all the way to Palm Beach and takes about 90 minutes for the 40 kilometre trip. You can also get buses 155 to 157 from Manly, but you will normally have to change at Narrabeen to the Palm Beach bus (190).

Some of Sydney's loveliest residential areas are tucked away in the Peninsula. Eighteen beaches are on the ocean side of the Peninsula and on the western side, at Pittwater, the beautiful, sheltered bays are lined with marinas and thick with the masts of sailing boats. In the summertime surf carnivals are held on

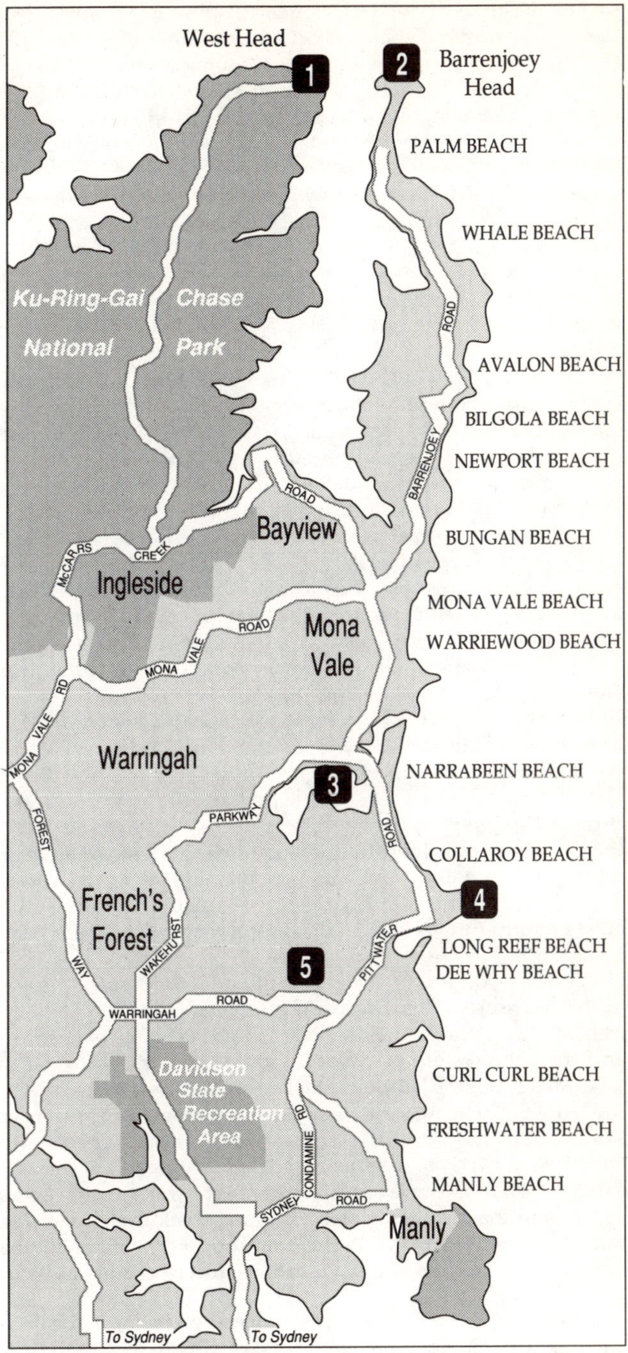

West Head
1
2
Barrenjoey Head
PALM BEACH
WHALE BEACH
Ku-Ring-Gai Chase National Park
ROAD
AVALON BEACH
BILGOLA BEACH
BARRENJOEY
NEWPORT BEACH
ROAD
Bayview
BUNGAN BEACH
McCARRS
CREEK
Ingleside
MONA VALE BEACH
ROAD
Mona Vale
WARRIEWOOD BEACH
MONA VALE
RD
MONA VALE
Warringah
3
NARRABEEN BEACH
PARKWAY
FOREST
ROAD
COLLAROY BEACH
French's Forest
4
PITTWATER
LONG REEF BEACH
DEE WHY BEACH
WAKEHURST
WAY
5
ROAD
WARRINGAH
Davidson State Recreation Area
CURL CURL BEACH
FRESHWATER BEACH
CONDAMINE RD
MANLY BEACH
SYDNEY
ROAD
Manly
To Sydney
To Sydney

1 West Head Lookout
2 Barrenjoey Lighthouse
3 Narrabeen Lagoon
4 Long Reef Lookout
5 Beacon Hill Lookout

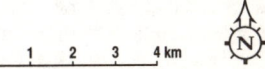

0 1 2 3 4 km
N

several of the beaches. It's a wonderful spectacle of surf lifesaving demonstrations, surf boat races, marching, etc. Usually the barbeques are churning out steak sandwiches and sausages by the hundreds for hungry spectators and participants. Ring the Surf Life Saving Association on 938 6944 to find out where a carnival is, or phone the tourist information centre at Manly on 977 1088.

Heading north from Manly, after the rocky headland at Queenscliff, the next beach is **Freshwater**. The walk from Manly is delightful. Otherwise, take buses 133, 134, or 137-39 from Manly. Like many of the beaches on the Peninsula, the beach front is largely undeveloped and unspoiled. The shallow water is perfect for swimming, body surfing and surfboard riding.

The next beaches, **Curl Curl** and **Dee Why** are both good surfing and swimming beaches. Buses 139 or 138 leave from Manly. The next headland north is Long Reef Point, and the beach to the south is **Long Reef Beach** which is popular with both wave jumping windsurfers and hang-gliders. Buses 155-157 leave from Manly, and routes 182-190 from Wynyard.

Collaroy and **Narrabeen beaches** form one long and beautiful stretch of sand. Buses 155 and 157 leave from Manly. Well north, and passing many beautiful beaches, but quite different in character, is **Whale Beach**. More secluded and intimate, the beach is framed with rocky headlands. Bus 190 goes from Wynyard to Barrenjoey Road near Whale Beach, but leave plenty of time for the trip and the walk over the hill.

At the extreme end of the Peninsula is **Palm Beach**. Bus 190 from Wynyard in the city will take about 90 minutes. There's plenty to keep you occupied on a day's trip to Palm Beach. Lunch cruises leave for the nearby beautiful waterways of Broken Bay and the Hawkesbury River. The Pittwater side of the Peninsula is wonderful for windsurfing. Rent windsurfers, motorboats or sailing dinghies here to enjoy the water. Though the community at Palm Beach is very small, there are several good restaurants, as well as the dining room of the Palm Beach RSL Club.

Whatever you find to do at Palm Beach, don't miss the walk to the **Barrenjoey lighthouse** on the headland. The sand spit leading to the headland has the crashing ocean surf on the east and the calm waters of Pittwater and the Hawkesbury on the west. From the top of the rocky headland, the views across to rugged West Head and the unspoiled Hawkesbury River are well worth the climb. For a real adventure, take the ferry from the Palm Beach wharf to the **Basin** or **Mackeral Beach** on the west side of Pittwater. There are no roads here, and the only access apart from the ferry is by steep walking track from West Head Road. Another lovely trip is the half-hour ferry ride to Patonga Beach on Broken Bay on Wednesdays and weekends at 9 a.m. and 11 a.m.

The marking of the **Warringah Coastal Walk** has made it easy to walk all along the coast of the Warringah Peninsula from Queenscliff Head near Manly to Barrenjoey Head. The track is 26 kilometres long and you almost certainly won't want to walk the whole track in one day. Why not stop overnight at one of the local hotels or motels and enjoy discovering a special part of Sydney? The walk is well marked and signs indicate whether the walk follows the road, walking trail, bushland track, or footpath. Highlights of the walk are the wonderful views from the many headlands. If you prefer a short walk, Narrabeen Head to

Turimetta Head will give the flavour of the coast. Further north, the steep climb up to Bangally Head gives wonderful cliff-top views from the highest point on the coast. Warringah Shire provides a pamphlet and map of the walk; phone 982 0333, or ask at the Manly Tourist Information Centre.

21   Bondi and the southern beaches

Bondi Beach

At break-neck speeds of 65 kilometres per hour, the steam tram from the city to Bondi Beach would hurtle down the final descent to the beach. In 1894, such speeds were horrifying to a society used to the leisurely pace of the horse. To "shoot through like a Bondi Tram" became a fond and enduring Australian expression. Sadly, the trams are long since gone and mostly forgotten. Instead, from the city, take the train on the Eastern Suburbs Line, to Bondi Junction. Buses 380, 381, 382 and 384 go to the beach from there. Or if you prefer, buses 380 and 382 leave from the Quay.

Bondi Beach has a place of special affection in Australian culture. Australia's most famous beach, Bondi is still evocative of hot sunny days on the beach, and bronzed lifesavers. In any other city in the world, Bondi Beach would be a special treasure. In Sydney though, many other beaches are just as beautiful. Bondi's attraction is something else. In the 1920s and 1930s, when Sydney was a far smaller city, the population was concentrated on the south side of the harbour, much of it in the eastern and nearby suburbs. Naturally, Bondi, only 7 kilometres from the city centre was far and away the most accessible ocean beach. The broad expanse of the beach and the low land behind the beach allowed Bondi to develop a beach front as many others could not. More recently, Bondi's racey image has helped it to become a cosmopolitan enclave.

To Australians spoilt by superb and unspoiled beaches, Bondi might seem ordinary and overcrowded. But if you are a visitor, you will surely love the broad expanse of sand, both wide and deep, good surf and safe swimming. Waikiki this is not! Framed by a tired and motley collection of apartment buildings and restaurants, a few decidedly seedy, Bondi Beach is being progressively smartened up. Yet the people of Sydney want Bondi much the way it is, not overwhelmed and overshadowed with exclusive high rises. Instead, a paint job, pastel colours naturally, and some sprucier new buildings are proposed to enliven Bondi and to regain some of its earlier glories.

Australia's first surf lifesaving club was established here in 1906. No need for lifesavers before this; a law passed in the 1830s prohibited daytime swimming in the harbour and the ocean from 6 a.m. to 8 p.m. Though rescinded around the turn of the century, the sight of exposed flesh continued to provoke distress and palpitations among the city fathers. "Morals saving" inspectors sternly patrolled the beach. Even in the 1950s, shortly after the bikini was invented, dedicated inspectors undertook the arduous and unenviable duty of measuring

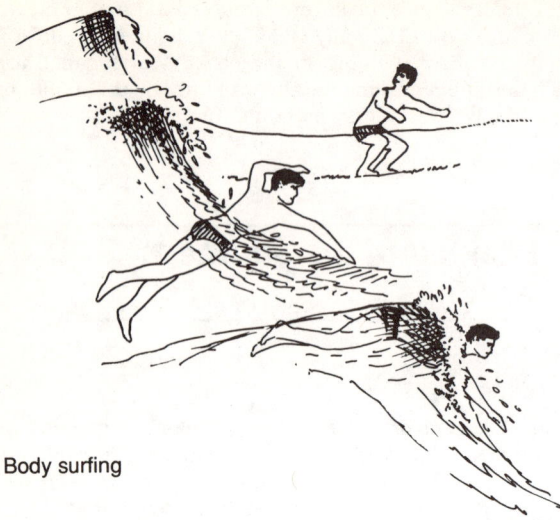

Body surfing

the length of women's bikini bottoms. Offending bikinis and bodies were escorted from the delicate public gaze. Times have certainly changed, topless bathing is now permitted, though restricted to the south end of the beach.

Swimmers should be careful to swim between the flags which mark the area of the beach that is safe from rip tides (the swift and strong currents which can drag weak and unsuspecting swimmers out to sea), and the part of the beach that is patrolled by lifesavers. Stay within the flagged area and you should have no problems, but if you do get into difficulties, just raise a hand high above your head. Most lifesavers rush to swimmers aid by fast rubber boats these days, so don't expect the full pageant of bronzed lifesavers.

One menace at Bondi and some other beaches is pollution. Unfortunately the sewage outfall is just a short distance north of the beach. When the swells drive the murk onshore, the beaches are closed.

A menace you don't have to worry about is shark attacks. The last unfortunate shark victim at a Sydney ocean beach was attacked in 1934. Bondi, like most of Sydney's beaches, has shark nets offshore. These have proved effective in discouraging sharks, even though they do not shut off the beach entirely. A shark patrol keeps careful watch, and ensures safety.

From the first of the early morning board riders and joggers to the last of the late night party pub patrons, something is always happening. Even if you don't fancy a swim Bondi is a fun place to picnic on the beach to watch the surfers and the beach life.

Whatever your tastes, there is no need to go hungry in Bondi. Choose from a wide range of restaurants, pubs, and discos. The Gelato Bar is famous for its delicious cakes, and the Bondi Fish shop serves good fish and chips in generous servings. There is also some reasonably priced accommodation near Bondi, including the backpacker's end of the market.

Beaches South of Bondi

Tamarama, the next beach south of Bondi, is a small but pretty beach, and very popular. Bus 391 leaves from Bondi Junction.

Water Quality at Sydney Beaches

Sydney, like most coastal cities, disposes its sewage waste into the ocean. When the sewerage from 3.5 million people is dumped into the water in a virtually untreated state it's not surprising that water can become polluted. There are three major ocean outfalls (Manly, North Bondi, and Malabar, south of Maroubra) dumping hundreds of millions of litres of effluents into the ocean each day. The water pollution problem is further aggravated by industry disposing toxic wastes into the water and through stormwater drains emptying into the sea after rainy weather.

Besides not being pleasant to swim in, there is evidence that the pollution causes ear and throat complaints as well as stinging eyes and possibly stomach upsets.

So where can you find clean water? As a general rule offshore winds will result in clean beaches: the sewage waste is lighter than sea water, floats on the surface, and is blown away by the off-shore winds. For the harbour beaches, the most important factor is rain. Under wet conditions the stormwater drains empty into the harbour resulting in poor water quality during and a day or so after the rain. Rain also affects the water quality at the northern and southern ocean beaches, particularly the northern surfing beaches and the Maroubra area.

To find out about water pollution conditions call **Beachwatch**, an independent water monitoring organisation, on 901 7996. Their recorded message will tell you whether pollution levels are low, medium, or high at several of the beaches and where to find the cleanest beaches. The message is updated with changing conditions.

So what is being done about the problem? Thanks to groups such as STOP (Stop the Ocean Pollution), formed in 1984, the awareness of water pollution has greatly increased in recent years. They helped sponsor a rock concert called "Turn Back the Tide" at Bondi Beach in 1989 which drew crowds of almost 250,000 to protest against sewage pollution. Another tremendous effort from the community has been organised by Ian Kiernan, a Sydney yacthsman. He has organised an annual clean-up to remove rubbish from the harbour and beaches. In 1990, 50,000 volunteers showed up on a Sunday in January and removed 5000 tonnes of garbage.

And finally, in 1989, the State Government made a commitment to do something about the pollution problem. Nick Greiner, Premier of New South Wales, promised that he would fix a "legacy of almost a century of neglect in terms of sewerage" and announced a $6 billion program to span 20 years. Unfortunately it may be some years before the water is significantly cleaner.

Bronte Beach has a sheltered swim area at the south end, as well as a pleasant park with picnic tables. Catch bus 378, the Bronte bus, from Bondi Junction. **Clovelly Beach** has calm, protected waters and is good for snorkeling. Bus 329, and the 390 Clovelly bus go from Bondi Junction. **Coogee** has quieter waters than the surf beaches and is good for body-surfing and swimming. Buses 313, 314 and 315 go to Coogee from Bondi Junction. **Maroubra Beach** is a large and popular surf beach, but has little of the charm of the others. Buses from the Quay include, 396 and 376 and 377, and from Railway Square, buses 393 and 395.

The southern most beaches in the Sydney area can be found at **Cronulla**, about 27 kilometres from the city centre. This is the only beach that can be reached by suburban train. In the summer there will be plenty of board riders on the train with their boards. It's a pleasant long beach with lots of activity, and when the sun becomes too hot, there is always an interesting walk along the shore.

Cronulla is a popular spot for fishing, with plenty of variety, and the whole estuary of the Hacking River to explore. At low tide the sand flats are ideal yabbie hunting grounds. From Gunnamatta wharves, just south of Cronulla Beach, tourist boats cruise Port Hacking and the Hacking River. The river marks the edge of the city. On the other side are a few little communities, Bundeena and Maianbar, and the bushland of the Royal National Park. A walk in the Park is strongly recommended. Check Part 6 for details. Ferries leave the wharf for Bundeena in Royal National Park every hour. One way tickets cost $1.80. Ring 523 2990 for times. To enjoy a more leisurely exploration of the river, why not rent a houseboat? Cronulla Houseboats rent houseboats to sleep ten people for about $700 for a weekend. Prices are often less during the week. The same company has fishing boats to rent for about $50 a day. Phone 523 6919 for more information.

22   Botany Bay

Now dominated by the airport and industry, Botany Bay was, in the 19th century, a favourite day excursion. A reserve on the southern shores has been left marking Captain Cook's landing place and some of the foreshore around La Perouse has been saved. On the western side of the bay, a long stretch of beach is backed by residential areas.

The Joseph Banks Hotel was a famous and unusual 19th century resort with its own zoo, circus, tents and toy train. Famous running races with world class professional runners were held here on a cinder track laid in front of the hotel. The grand New Brighton hotel at Brighton-Le-Sands was another jewel on Botany Bay and opened in 1886 complete with a ballroom, skating rink and sea baths.

A cruising restaurant explores Botany Bay and the Georges River—an interesting way to see some of Sydney's waterfront suburbia. Boat departs daily from Sans Souci. Phone Bass, or Flinders Riverboat Cruises on 526 6188.

Captain Cook's Landing Place

If you are travelling by car, take the Princes Highway south to Kogarah and turn onto Rocky Point Road. Rocky Point Road becomes Taren Point Road. From here take Captain Cook Drive to the end of the peninsula. By public transport, take the train to Cronulla Station, and then Kurnell Bus 67. The landing place

*is open from 7.30 a.m. to 7 p.m., and the entrance fee is $3.50
per car.*

Captain James Cook, in 1770, was the first to map the east
coast of Australia. On an expedition to search for the fabled
southern continent, his first landing point was here at Botany
Bay on 29 April 1770. He first chose the name Stingray's
Harbour because of the large numbers of stingrays he saw in
this wide, shallow bay. He later renamed it Botany Bay in
recognition of the great quantity of unusual plants found by his
botanist, Joseph Banks.

A large stone pillar marks the place where Cook and his crew
came ashore. The fascinating story of Captain Cook and the
voyage of the his ship the **Endeavour** is told in the small
museum. A prize exhibit is one of the cannons from the
Endeavour which was salvaged on a reef in Northern Australia
in 1969. The Endeavour had become stuck on the reef and the
cannons were thrown off to lighten the load. A model of Endeavour is displayed, together with bits and pieces from the
ship.

The landing area and surrounding reserve make for an
interesting picnic spot, with barbecues, tables and hiking trails
around the reserve. However the nearby oil refineries at Kurnell
definitely take away some of the atmosphere of this historic
landing place.

La Perouse

*La Perouse is on the northern side of Botany Bay at the end of
Anzac Parade. By bus, take the 393 from Railway Square or 394
from Circular Quay or Elizabeth Street.*

The great French explorer, Comte de La Perouse landed here
in 1788, only a few days after Captain Phillip arrived with the
First Fleet. The French came to this quiet bay to recuperate
after a attack by islanders in Samoa. Bare Island Fort is on a
tiny island joined to the mainland by a bridge. Built in 1885,
the fort formed part of the colony's chain of defence works. The
Macquarie Watchtower dates from the 1820s, and was built
to keep track of ships entering Botany Bay. The **La Perouse
Museum** traces the explorer's fascinating history, showing
original maps of his voyage and navigational instruments.

Two scenic nature trails at La Perouse along the coast pass
by the 19th century forts of Henry Head and Cape Banks, and
the shipwreck of the **Minmi**. An added attraction at La Perouse
is the Snake Show in The Snake Pit here every Sunday. John
Cann runs the show aimed at educating the public about
snakes, rather than displaying feats of bravado. The show has
been running since early in the century. A couple of locals
usually come down to La Perouse on weekends to sell their
hand-carved boomerangs—they also give throwing demonstrations.

PART VI
Exploring Sydney's Bushland

23   The natural environment

There is no disputing Sydney's natural beauty, in the same
league as say San Francisco, Hong Kong, and Rio de Janeiro
as the most beautiful harbour cities in the world. Sydney is
Australia's largest city with 3.5 million people. The urban area
covers some 1500 square kilometres, a vast area spreading
along the coast and almost into the Blue Mountains.

The eastern parts of Sydney are dominated by the harbour
and its intricate waterways. The harbour and waterways combined with the undulating topography of a series of sandstone
ridges has greatly influenced the pattern of development. A grid
system of roads would have been very difficult and impractical
on such a landscape. In the west The Cumberland Plain is
largely flat and extends to the rugged Blue Mountains. Three
rivers cut through the Sydney area: the Hawkesbury River,
which is navigable for a stretch of 110 kilometres inland, the
Parramatta River, and the Georges River in the south.

Flora and Fauna

The soils characteristic to most of the Sydney Region are derived
from sandstone and consequently are sandy, do not hold water
well and are low in fertility. The infertility of the soils is one of
the reasons why much of the land that is now protected in the
national parks was not cleared for farming. The early settlers
found the fertile soils they needed along the Hawkesbury River.
Despite the harsh conditions there is an enormous variety of
native plant life, many of the plants having hard, tough and
small leaves. This is because of a lack of nutrients in the soil
and protection against drought and heat.

A visitor to Botany Bay in the 1880s, perhaps expecting a
rainforest after Joseph Bank's excitement over the tremendous
variety of plants remarked:

> Our opinion was, that he (Joseph Banks, botanist on Captain
> Cook's ship) had collected them far too carefully, and left
> none remaining, for it appeared to be the most desolate and
> barren spot we had seen all around the city. (J.A. Ure, *A Tour
> Around the World*, 1885)

Native plants on the headlands and cliff tops are generally
heath, mainly low-growing due to the shallow soils and ex-
posure to wind and sun. The strange banksia tree also grows
in these heath areas. In the woodland areas of Sydney the
dominant vegetation are Eucalypts (gum trees). Over 500
species of gum trees are found throughout Australia. In some
protected gullies, pockets of rainforest can still be found in the
Sydney area.

One of the most beautiful trees found in and around Sydney
is the fig tree. They are very large trees with shiny green leaves,
huge spreading boughs, and massive smooth trunk. The
Moreton Bay fig from Queensland is the largest. The Port
Jackson fig, indigenous only to the Sydney area, has smaller
leaves. The fig trees forming the avenue in Hyde Park are Hill's
fig, a variety from Queensland.

Sydney's mild climate enables a tremendous variety of introduced plants to thrive. In the spring the beautiful bougainvillea
bush brightens up many suburbs with its splash of vibrant

Sydney Harbour Bridge and Opera House (*Viewslides*)

The great ferry boat race

Bushwalking
in the Blue
Mountains
(*Tourism
Australia*)

A hire yacht on the beautiful Hawkesbury River
(*Pittwater Yacht Charter*)

Hot air ballooning in the Hunter Valley (*Tourism Australia*)

Horse riding in Centennial Park (*Northside Productions*)

Skiing in the Snowy Mountains (*Tourism Australia*)

Speaker's Corner, the Domain (*Northside Productions*)

Colour and movement—Sydney's Gay Mardi Gras
(*Northside Productions*)

colour. The Jacaranda tree blossoms in November, with a delicate purple flower. Hibiscus, poinsettias, and oleanders also contribute to the colour. No matter the time of year most gardens have some flowering plants. Citrus trees grow well in Sydney and even the occasional banana tree can be spotted in a Sydney garden.

Flying foxes (fruit bats) are surprisingly common in the Sydney area. They can often be spotted in fig trees going after the purple fruit, especially around dusk. Possums are fairly common and come out at night looking for food and often rummage through garbage. Kangaroos of course once roamed the harbour shores but now are generally only found in Royal and Kur-ing-gai National Parks. The best chance of seeing them is around dusk. The birdlife in Sydney is rich and varied. Crimson rosellas, rainbow lorikeets, and sulphur-crested white cockatoos are among the most colourful. Many Sydneysiders still wake up to the laugh of the kookaburra. The magpie is a distinctive large black and white bird. And you're sure to see the introduced mynah bird. This cheeky intruder seems to be everywhere: brown coloured and unmistakable with its black head and yellow beak and legs. In the parks, seagulls will always attempt to join in your picnic and you may even have an ibis poking around with its long beak.

24   Parks

There are lots of beautiful parks and gardens around Sydney. Many of them are described as part of the suburbs in which they are located. This section provides a quick summary of the outstanding parks and gardens as well as details on the national parks surrounding Sydney. Check the index for additional information on the parks in other sections of this book. An excellent source of information about National Parks and the natural environment is the National Parks and Wildlife Shop in the Rocks at Cadmans Cottage, 110 George Street. They have a wide range of free pamphlets and maps. They also sell maps and interesting publications and posters. Open every day from 9 a.m. to 5 p.m., phone 278 861. The Service runs an Information Line, Monday to Friday between 8.30 a.m. and 4.30 p.m., on 585 6333.

Inner Area Parks

The **Royal Botanic Gardens**, the **Domain**, as well as **Hyde Park** form a green belt on the east side of the downtown. It's a great bonus for Sydney to have such an extensive and beautiful patch of green so close to the centre. They make the perfect refuge for those seeking a quiet picnic or just a place to sit down away from the hustle and bustle of the city. See Part 3 for a detailed description.

At **Watsons Bay** there is a pretty little park with magnificent fig trees at the water's edge and while you're here don't miss the park and a swimming beach nearby at Camp Cove. There is a dramatic clifftop walk at the Gap across the street from the Watsons Bay Park.

Neilson Park on Greycliffe Avenue in Vaucluse (Bus No. 328) is a lovely shady park and has a swimming area with sharkproof nets in the summer. **Ashton Park** (Taronga Zoo ferry) is a large park with lots of woodlands, pleasant walks, wonderful views, and picnic grounds. **Dobroyd Head** and **Grotto Point** are situated between North Head and Middle Harbour. The panoramic views of Sydney and the harbour are stunning and at Grotto Point there is a delightful lighthouse.

Centennial Park, Oxford Street, Paddington. This huge expanse of green smack in the middle of the eastern suburbs is a popular place but there's plenty of room for everyone. Bicycle rentals and horse hire available, see Part 8.

Blues Point, the park at the end of McMahons Point, gives wonderful views of the harbour. If you're after shade, go up beyond the tall apartment block on the point. On the west side of the point, there is Sawmiller's Reserve. (Take Hegarty's ferry to Blues Point.)

Lavender Bay is a shady place with views across the bay and towards the city. (Take Hegarty's ferry to Lavender Bay.)

The tunnel construction is going to disrupt **Bradfield Park** until 1992, but if it's a picture of the skyline and bridge or Opera House you want, there's still room to click. (Milsons Point Railway Station and Hegarty's ferry to Jeffrey Street wharf.)

Cremorne Point is another lovely place with good views of the harbour and most enjoyable walking paths on either side of the point. (Ferry to Cremorne Point)

Balmoral Beach, Middle Harbour, is a stretch of harbour beach which has plenty of shady areas, provided by big fig trees right behind the beach. It's a popular place on summer weekends, so parking can be difficult.

Long Nose Point Reserve, Parramatta River, Louisa Road, Birchgrove is a popular place for fishing from the wharf or stone walls.

Suburban Parks

Lane Cove River State Recreation Area, Lady Game Drive Chatswood. Only a few kilometres from the city centre, this is a very popular park. It has extensive picnic areas, a children's zoo, plenty of ducks to feed. Rowboats can be hired or you can even take a ride on a paddle-wheel steamer down the Lane Cove River.

The **Davidson Park State Recreation Area** covers the bushland around the upper reaches of Middle Harbour. Activities include walking, boating, canoeing, fishing, and horseback-riding. There are pleasant, shady grassed areas along the shores of Middle Harbour for picnicking. The walking trails lead through gullies, along streams, the foreshores, and through gum forests. Detailed maps are available at the park entrance on weekends. Phone 451 3479. Car entrance fee on Sundays, $4 and boat launching fee is $5 on weekends. Access road is just east of Roseville Bridge, off Warringah Road (12 kilometres from central Sydney).

Bicentennial Park, Australia Avenue, Concord West, was created to celebrate the Bicentennial in 1988. It has timber boardwalks over the mangroves thickets on the Parramatta River. The mangrove areas and waterbird refuge offer good birdwatching. There's a lookout tower for views of the river and the Sydney skyline. For pedestrians and cyclists a number of paths twist their way around the park. The Wisteria Tea House provides light refreshments and overlooks the lake. There are also very good picnic facilities. Open daily sunrise to sunset. Phone 763 1844. Concord West railway station is a 5 minute walk to the park.

Parramatta Park, Parramatta, is a very pleasant large old park on the shores of the Parramatta River. See Part 4.

Botany Bay National Park takes in coastal areas on the north and southside of the bay. At Captain Cook's Landing Place on the southern head of Botany Bay there are picnic grounds at the water's edge, walking trails, and the Captain Cook Museum. The north side of the Bay at La Perouse has another good picnic spot and walking trails as well as La Perouse Museum.

Georges River State Recreation Area, Henry Lawson Drive, Georges River, has several access points along the shores of the Georges River. Burrawang Reach and Fitzpatrick at Picnic Point has fishing, bushwalking, and picnicking. Oatley Park on the Georges River is another popular spot and has a swimming enclosure, barbecues (with wood supplied), and nature walks.

25   Gardens

Bankstown Native Garden, 7 Sylvan Grove, Picnic Point: all the plants along the track are marked, so it's a very pleasant way to learn about native plants in this lovely location near the Georges River. There are over 300 species of native plants, including a rainforest area. Bus from Revesby Station. Open weekdays 7 a.m. to 3 p.m. and weekends during the spring (mid-August to November). No admission fee (771 1650).

The **Stoney Range Flora Reserve** is a little reserve with spectacular wildflowers in the spring (end of August to mid-October) and a lovely, little rainforest section. Pittwater Road, Dee Why, buses from Wynyard or Manly. Open every day 10 a.m. to 4 p.m. Admission 50 cents (982 9314).

Ku-ring-gai Wildflower Garden, 25 kilometres north of the city on Mona Vale Road in St Ives. Lovely gardens, mostly natural bush, and an enormous variety of Australian plants. Buses from Pymble Station. Open daily from 10 a.m. to 4 p.m. (440 8609).

E.G. Waterhouse National Camellia Garden, President Avenue, Caringbah, about 1 kilometre from Caringbah railway station. This is a must for camellia fans, with over 1500 varieties. The camellia season is late June to mid-October. (524 6090)

Royal Botanic Gardens: see Part 3 for a description. The Royal Botanic Gardens have established two additional gardens:

Mount Tomah Botanic Gardens, about 100 kilometres west of Sydney on the Bells Line of Road, Blue Mountains. Cool climate gardens featuring plants of the Southern Hemisphere, located at 1000 metres above sea level with sweeping views of the mountains. Open 10.30 a.m. to 4 p.m. March to September, and 10.30 a.m. to 6.00 p.m. October to February. Admission $5 per car. (045 67 2154)

Mount Annan Botanic Gardens, about 60 kilometres south-west of Sydney at Campbelltown, off Narellan Road. This recently established garden specialises in Australian native plants. The gardens are a massive 400 hectares, with an 8 kilometre scenic drive, 20 kilometres of walking trails, picnic grounds, and thousands of plant species from all over Australia. Among their vast collection are 500 different species of gum trees; about 25 species of bottlebrush are displayed, and over 70 different banksias. Buses run between Campbelltown railway station and Camden, via Mount Annan. Ring Macarthur Coaches for a timetable (046-66 7501). The Gardens are open 10 a.m. to 4 p.m. from April to September, and 10 a.m. to 6 p.m. from October to March. Admission $5 per car (046-46 2477).

26   The bush around Sydney

Ku-ring-gai National Park

A number of roads lead into the park: the Bobbin Head area is reached via Pymble on the Pacific Highway; from the Terrey Hills Road, a road to the the east to Church Point; a road to the west following Coal and Candle Creek and to the north all the way up to the spectacular West Head. By public transport you can reach some hiking trails in the Bobbin Head area (see below) and Deane's Bus, Route 223 goes to the Chase Gates, Bobbin Head. Ring 888 3022 for bus times. Buses go to Church Point and there are ferries from Church Point and from Palm Beach.

This beautiful park just 25 kilometres north of Sydney was designated a national park as early as 1894. Many waterways run through the park, which is bounded on the east by Pittwater. This Hawkesbury sandstone country produces magnificent wildflowers and over 900 different plants have been recorded. Winter and spring are the best times to view the wildflowers in the park.

Highlights of the park

A drive to the top of the park takes you to the spectacular **West Head**. From this point you get an unbelievable view of the surrounding waterways, across Broken Bay and Pittwater. You'll see Lion Island, its shape resembling a crouched lion, and if it's a fine weekend, hundreds of boats will be sharing the waters. Don't forget your binoculars and camera.

Another lovely drive takes you along the dramatic **Coal and Candle Creek**, a large waterway with steep wooded shores. This road passes Akuna Bay where you'll see a multi-storey boat park. It's a nice place to stop for an ice-cream and watch the boat activity.

Many of the West Head walks will take you down fairly steep **walking tracks** to quiet beaches. The short walks from the picnic areas are well worth the walk for the views and the interesting plant life.

Located on the Ku-ring-gai Chase Road between the Bobbin Head Picnic Area and Mt Colah is the informative **Bobbin Head and Kalkari Visitor Centre**. It has a nature trail; the names of trees and plants are marked, and you'll see kangaroos, emus, and lots of birdlife. The centre is open from 9 a.m. to 4.15 p.m. every day, phone 457 9853.

At Bobbin Head there are extensive picnic grounds as well as fishing and boating spots. Good hiking trails are found in this area, two of which are accessible by public transport. You can reach the **Berowra track** from Berowra station and the **Mt Ku-ring-gai track** from the Mt Ku-ring-gai station.

The Basin is a popular camping spot and can be reached by walking or by ferry from Palm Beach. Bookings are essential for camping. Ring 919 4036 between 9.30 a.m. and 10.30 a.m on weekdays.

There is another nice drive leading to **Church Point**. A ferry leaves from Church Point up to the Pittwater Youth Hostel at Halls Wharf. Phone 99 2196 for hostel bookings and information on how to get there. This ferry also calls at other points on Pittwater and is a delightful ride. You can get to Church Point by bus from the city, and the ferry leaves about every hour on weekdays and half-hourly on weekends. There is a grocery shop at Church Point.

Brisbane Water National Park

Brisbane Water is on the north side of the Hawkesbury. The Wiseman's Ferry Road, Girrakool, by the Pacific Highway, and Patonga Drive, from Woy Woy, all give access to the park. Walkers can get off at Wondabyne railway station or take the ferry from Brooklyn or Palm Beach to Patonga Beach.

Like Ku-ring-gai, the common rock type in the park is sandstone, and it also has an enormous variety of wildflowers. It has several interesting bushwalks, some offering spectacular views over the waterways. A walkway is provided to view Aboriginal carvings in the flat, exposed sandstone areas located at the **Balgandry Aboriginal Engraving Site** on Bambarra Road, north of Woy Woy. For further information phone the National Parks and Wildlife Service on (043) 24 4911 (Gosford office).

Bouddi National Park

This park is situated along the coast on the northern entrance to Broken Bay. It is about 20 kilometres south-east of Gosford on the Scenic Road through Kincumber or via the Empire Bay Road and Wards Hill Road from Ettalong.

For quiet beaches and pristine waters this is the place to go. If you're willing to make a short walk away from the car access points, you'll have the beach to yourself. Car camping is permitted at Putty Beach as well as a few places for bush camping. All campers are required to make an advance booking by calling 043 244 911. Mount Bouddi has a picnic ground.

Royal National Park

The park is 36 kilometres south of Sydney by road, via the Princes Highway. The two access points from the highway are Farnell Avenue south of Loftus, and McKell Avenue at Waterfall. By public transport you can get off at several stations such as Loftus, Engadine, and Waterfall. A weekend service stops at the Royal National Park station at Audley. A ferry service leaves

Cronulla wharf every hour on the half hour to Bundeena and this is a popular access point with bushwalkers. For more information on the ferry service, phone 523 2990.

Royal National Park was the first area in Australia set aside as a National Park (in 1879), and the second in the world, after Yellowstone National Park in the United States. The park can suit all moods—from a quiet paddle in the river and creeks to lengthy bush walks. The coast is incredibly varied, and has both scenic rugged cliffs, some soaring to 300 metres high, and good surf and swimming beaches.

Swimming

You can reach several beaches by car, such as Bonnie Vale, Wattamolla, and Garie Beach. If you're willing to walk a short way from the car park you can have beaches to yourself.

Bushwalking

There are dozens of interesting trails to pick from, ranging from easy strolls to overnight walks. The coast track extends the length of the park, where you can walk along cliff tops, see waterfalls and caves, go through palm jungles and pass irresistible beaches. For the adventurous, it's possible to do a two-day walk from Bundeena all the way south to Otford, a distance of 26 kilometres. There's a small youth hostel located at Garie, about half way to Otford. Otherwise, camping is permitted almost anywhere with a permit. The **visitor centre** on Ironbark Flat at Audley has free pamphlets and maps of the trails.

Birdwatching

The park is popular with birdwatchers, and the best areas for viewing are Bonnie Vale or Scientist Cabin. Over 200 species of birds have been spotted in the park.

Canoeing and Rowing

Rowing boats and canoes are available for hire at Audley.

Blue Mountains National Park

About 100 kilometres west of Sydney, the Great Western Highway through the Blue Mountains provides many access points to this enormous park which covers 200,000 hectares. There is a frequent train service to the mountains from Central Station and choice bushwalking areas in the park can be reached from Blackheath railway station.

Although the Blue Mountains are not a high mountain range, the scenery is dramatic, with deep gorges, waterfalls, and high cliffs offering wonderful views. The vegetation is mostly gum forest but there are also areas of rainforest and heath lands. These forests support diverse wildlife including lyrebirds, honeyeaters, crimson rosellas, cockatoos, wallabies, and possums.

The information centres have an excellent range of free pamphlets on the activities, facilities, and maps of short and long walks. The **Glenbrook Visitors Centre** is in the Lower Blue Mountains on Bruce Road, Glenbrook. It is open weekends only, phone 047 392 950. At Blackheath there is the **Blue Mountains Heritage Centre**, on Govetts Leap Road. It is open every day from 9 a.m. to 4.30 p.m., ring (047) 87 8877.

Bushwalking

The mountains have an enormous variety of interesting bush-walks, ranging from short strolls to the cliff's edge to strenuous several day's walks. A note of caution—the weather can change quickly so always bring cool weather clothing and a rainproof jacket. On most tracks there is no freshwater available, so you will have to carry your own. For maps and information about trails, contact the information centres listed above. There are also several good bushwalking guidebooks available.

Short walks

The **Red Hands Cave walk** is about 3.5 kilometres (3 hours return), beginning at the Glenbrook Visitor Centre. It leads to a cave which was once inhabited by Aborigines. The cave's name comes from the red ochre hand stencils decorating the walls of the cave. Several other short walks start from the Glenbrook Visitor Centre.

The **Gorge walking track**, in the Lower Blue Mountains, on the Bell's Line of Road at the Waratah Native Gardens, is an easy one hour walk. It takes you through gum forest and then down to rainforest in the gorge. The unusual Gang Gang Cockatoo with its red head and crest and grey body is often spotted along this trail.

Around **Wentworth Falls** several walking trails take you in, around, or even under the dramatic falls. In the **Blackheath** area, pick from several short and long walks. Some of the short walks have very steep descents. One of the most pleasant ones without steep grades is the **Govetts Leap to Evans Lookout Clifftop walk** which has lovely long distance views and takes about 3 hours return. The **Fairfax Heritage walk** from the information centre at Blackheath is about 2 kilometres long and very easy. It offers views over Grose Valley and has inter-pretive signs as well as a self-guided walk pamphlet.

Long walks

At Blackheath there are a few full-day, fairly strenuous walks. Call in at the information centre to get advice on long walks. The **Govett's Leap to Perry's Lookdown walk** takes you through the lovely Blue Gum Forest and down to the Grose River. Another good walk in this area is the **Evans Lookout to Grand Canyon and Neates Glen**. At the bottom of the canyon there is lush rainforest and lots of little waterfalls.

The **Ruined Castle, Mount Solitary walk** will take you through the country you look out over at Echo Point. The walk begins at the top of "The Golden Stairs", off Cliff Drive and down Narrow Neck Peninsula. It takes about 5 to 6 hours return.

The **Lockley Pylon walk** out of Leura is an exceptional walk in the Blue Mountains. This is one of the few walks that gives expansive views, walking through low heath rather than forest. The access to the beginning of the walk is on Mount Hay Road out of Leura. The road is unsurfaced and should not be attempted in wet weather without four wheel drive. The walking track begins about 9 kilometres down Mount Hay Road at the Fortress Ridge Jeep Track. The walk is 9 kilometres and there are no steep grades.

Camping

Camping areas are provided at Perry's Lookdown, at the end of Hat Hill Road at Blackheath and at a few spots near Wentworth Falls. Some bush camping sites are also provided for walkers.

---

27   Australian wildlife

Besides Taronga Zoo (described in Part 5) there are quite a few places to observe Australia's fascinating wildlife—some in their natural habitat!

Aquariums and Ocean Fauna

Sydney Aquarium at the eastern entrance to Darling Harbour. The Sydney Aquarium is the largest in the world. The highlights of this fascinating attraction are the two enormous oceanaria. Visitors walk though acrylic tubes: this could be the closest you'll get to walking on the ocean floor without scuba gear. One oceanarium contains marine life from Sydney Harbour and the other replicates the open sea where you'll find youself face to face with two metre grey nurse sharks or cowtail rays. Open 9.30 a.m. to 9 p.m. every day, admission $10 adults, $5 for children, and senior citizens, $7. (262 2300)

Manly's Underwater World at the harbour beach at Manly. Watch sharks being fed, or take a walk through the sea on a moving footway. There's a seafood restaurant here and the Manly Waterworks, with its four giant waterslides. Open daily 10 a.m. to 5 p.m. Admission $10 for adults, $5 children. (949 2644)

On the Open Sea

How would you like to explore the ocean wilderness south of Sydney in search of whales and dolphins? The "Off the Shelf" Marine Wildlife Survey company takes passengers along on their regular surveys. Humpback whale migrations are normally from June to November and dolphins are seen all year round. Many kinds of birds and fish are often spotted. A hydrophone is lowered in the water to record the sounds that may be heard. Remember this is not just a quiet harbour cruise, the open sea can get rough and of course no guarantees can be made about what you'll spot.

Boats leave from Rose Bay. At some times of the year they leave from Wollongong Harbour which is about a 2 hour train ride fron Central Station. The boat leaves at 9 a.m. and returns about 3 p.m. The cost is $40 for adults and $35 for children. Tea and coffee are provided free but you'll need to pack a lunch. Ring 528 2505 to book a trip, or write to Off The Shelf, PO BOX 90, Oyster Bay, NSW 2225.

Birdwatching

There are over 700 species of native birds in Australia. In the

Sydney area alone over 400 species have been spotted, including some non-native species. The best months for birdwatching are September, October, and November. Right in the heart of Sydney the chances of seeing a fabulous rainbow lorikeet or a sulphur-crested cockatoo are good, especially along the foreshore walks of Cremorne and Mosman. The Royal Botanic Gardens, Centennial Park and the Taronga Zoo grounds are good inner city locations but of course if you wander further afield your chances of seeing a larger variety are much greater.

Ku-ring-gai National Park, north of Sydney is a good place to go. At West Head the birdwatching is good and the views are spectacular. Long Reef at Collaroy is also a good spot. Royal National Park on the south side is another choice place and there are weekend trains to Royal National Park station at Audley or you can take a very early morning ferry from Cronulla wharf to Bundeena and head off to Bonnie Vale in the park. See sections above for access to these parks. The Hawkesbury River and nearby swamps are good spots as well as the road north from Wiseman's Ferry. Cumberland State Forest on Castle Hill Road in the north-west is a favourite place for birdwatchers. Barrengrounds Nature Reserve, Jamberoo, south of Sydney is famous for its birdwatching and the Royal Australasian Ornithologists Union runs an observatory here. Ring the warden on (042) 36 0195.

The Australian Museum bookshop at College and William Street has a wide choice in field guides and a bird display complete with bird sounds. The contact numbers for birdwatching clubs can be obtained by ringing the Environment Centre of New South Wales on 27 4206. The New South Wales Field Ornithologists meet the first Tuesday of every month in the Australian Museum and visitors are welcome.

Koalas and Kangaroos

Koala Park, Castle Hill Road, West Pennant Hills. Located in the north-west of the city, it is about 30 kilometres from the city centre. By public transport take the train to Pennant Hills station (about 45 minutes) and then bus No. 655 to the Koala Park. Here's your chance to cuddle a koala but count on taking a turn with dozens of other koala lovers. There are also kangaroos, wombats, cockatoos and a picnic area. Best koala viewing times are 10.20 a.m., 11.45 a.m., 2 p.m., 3 p.m. Open daily 9 a.m. to 5 p.m. (875 2777).

Waratah Park, Namba Road, Duffys Forest near Terrey Hills. North of Sydney, nestled in a corner just outside the national park, it is about 35 kilometres from the city centre. By bus, the Forest Coach Lines Bus No. 56 to Duffy's Forest departs Chatswood railway station at 11.05 a.m. on weekdays. For a bus timetable phone 450 1236. This park has a lovely setting overlooking Ku-ring-gai National Park and the Hawkesbury River in the distance. You can pet koalas and kangaroos, and you might see wombats, dingos, and emus. They have a little bush railway going on a 15-minute ride down a shady gully. A bush barbeque lunch with cook-your-own steaks is available as well as picnic areas. Open daily 10 a.m. to 5 p.m. (450 2377).

Featherdale Wildlife Park, Kildare Road, Doonside, just west of Blacktown, about 30 kilometres from the city centre. By public transport, take the train to Blacktown Station, then Bus No. 211. There is a picnic area and kiosk, and a wide variey of native birds as well as reptiles, Tasmanian devils, and other native animals. The koala enclosure opens at 10.30 a.m. and 2.30 p.m. The Park opens daily from 9 a.m. to 5 p.m. (622 1644)

Sheep Shearing

If you want to see a sheep being shorn the Aussie way, there's the **Sydney Agrodome** the Dalgety Centre, RAS Showground, Driver Ave, Paddington. Buses from Circular Quay and ask for Driver Avenue entrance to the Showground. After shearing, there are demonstrations on how the wool is turned into the final product. There is a craft shop selling wool products and a restaurant serving lunch. Open daily 8 a.m. to 6 p.m. Three shows a day at 9.30 a.m., 12 noon, and 2.30 p.m. Admission is $9 for adults, including Aussie tea, and $4.50 for children (331 7279). Another place to watch sheep and shearers in action is north-west of Sydney in Maroota at the **Tobruk Merino Sheep Station**, phone 954 4566.

Other Wildlife

About one hour's drive west of Sydney at Warragamba Dam is the **African Lion Safari Park** with tigers and bears, etc. A 5-kilometre drive through the park allows you to view the animals from your car. There is also a sea lion show, parrot circus, kangaroo reserve and a safari railway. Open Wednesday to Sunday from 10 a.m. to 4.30 p.m. (047-74 1113).

PART VII
Beyond Sydney

28   Windsor and the Upper Hawkesbury River

Windsor is 56 kilometres north-west of Sydney and Richmond is a few kilometres beyond. There is a regular train service to Richmond and Windsor.

Settled from the early years of the colony, the rich fertile soil combined with the navigable Hawkesbury River made it an ideal place for agricultural production. Land grants were taken up as early as 1794 and in 1810 Governor Macquarie established five new towns in this area: Pitt Town, Wilberforce, Castlereagh, Richmond and Windsor. Richmond and Windsor are the only towns that have remained as commercial centres but a tour of all of the towns makes for a fascinating drive.

The Department of Planning (175 Liverpool Street) has prepared a detailed and beautifully illustrated walking tour of Richmond and Windsor; it costs $1.00.

Windsor

Windsor has a population of about 6000 and was the most important of the five Macquarie towns. It was the administrative centre and the transport hub for agricultural produce being carried by river boats to Sydney. It's a pretty town with many lovely colonial buildings remaining. Laid out according to the original plans, the centre is the traditional town square, called Thompson Square. It is surrounded by stately 19th-century buildings and fronts the bank of the Hawkesbury River. The Museum and Information Centre are on this square and here you can get maps of the town. The centre is open from 10 a.m. to 4 p.m. weekdays and 10 a.m. to 5 p.m. on weekends.

The buildings on either side of the tourist centre have a long history. The **Macquarie Arms** was built as as inn in 1815. The wide porch provides a convenient spot for a cool drink on a hot summer day. The **Doctor's House** on the other side of the centre is an elegant Regency brick terrace. Macquarie's architect, Francis Greenway, designed two lovely buildings here. The **Court House** on Court Street was built in 1822 and is still used as a court. **St Matthew's Church of England** at the other end of town near McQuade Park was completed in 1823. Across from the church is the pretty **St Matthew's Rectory** built in the late 1830s by William Cox. The church has a lovely positon near the top of the hill and overlooks the town and river plains. The cemetery has graves dating from as early as 1810.

Richmond

Richmond has a population of about 8000 people. This is another of Macquarie's planned towns and has a lovely village green in the centre. The **Blackhorse Inn** on Bosworth Street was built in 1819, conviently located near the finish line of a street horse racing track. No longer a pub, it is now used as an apartment building.

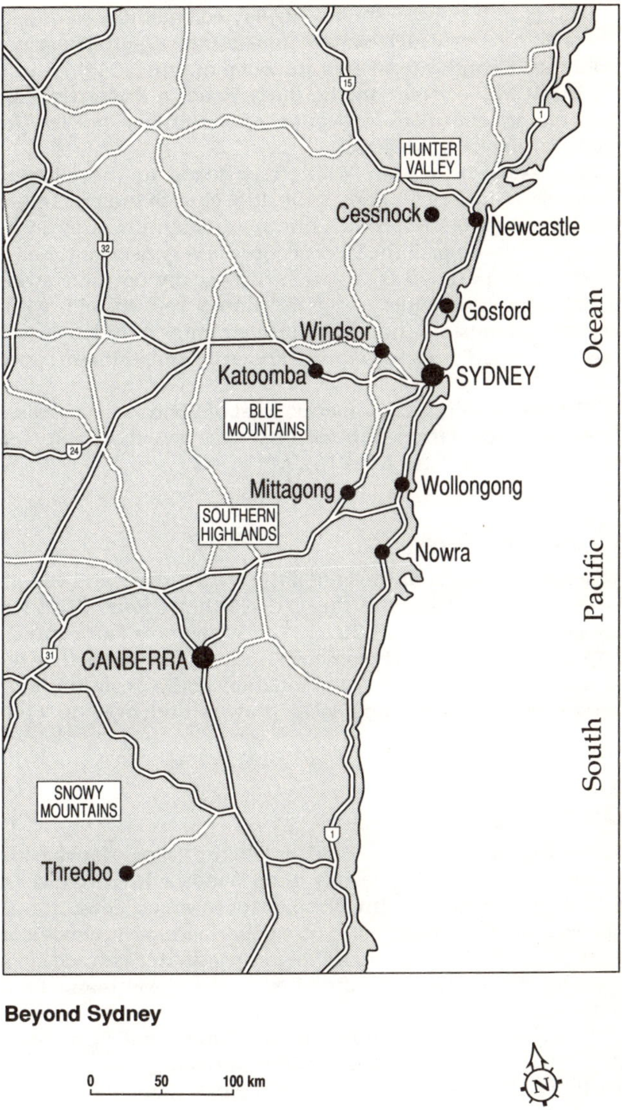

HUNTER
VALLEY

Cessnock ● ● Newcastle

Gosford ●

Windsor ●

Katoomba ● ● SYDNEY

BLUE
MOUNTAINS

Mittagong ● ● Wollongong

SOUTHERN
HIGHLANDS

Nowra ●

CANBERRA ●

SNOWY
MOUNTAINS

Thredbo ●

Ocean

Pacific

South

Beyond Sydney

0   50   100 km

Wilberforce, Ebenezer and Pitt Town

At Wilberforce, about 6 kilometres from Windsor there is an attractive reconstructed village called **Australiana Pioneer Village**. Many authentic buildings have been moved here from the countryside. The shops, a church, a bank, a school with a museum, and railway station display colonial day-to-day life. There is an inn which serves Aussie fare. Open Tuesday to Friday and Sunday from 10 a.m. to 5 p.m.; ring (045) 75 1457.

Also at Wilberforce, on the Putty Road, a **Butterfly Farm** features an enormous collection of butterflies, beetles, and spiders; ring (045) 75 1955.

The road north from Wilberforce leads to the village of **Ebenezer**. The delight here is the little church built in 1809. It was the first Prebysterian church in Australia. Continuing along this road north the river scenery is very pleasant. Nearby on the Tizzana Road is "a touch of Tuscany on the Hawkesbury". This little winery is more than a century old, with a beautiful sandstone building situated on a secluded twist in the river. Open weekends or weekdays by appointment, phone 045 79 1150.

Pitt Town, about 7 kilometres west of Windsor is a pleasant tiny village. **St James Schoolhouse** was constructed in 1820 and the church dates from 1859.

Wiseman's Ferry

Wiseman's Ferry is at the end of the Cattai Road, and can also be reached by following the Old Northern Road. The ferry crosses the Hawkesbury River. The Wiseman's Ferry Inn was originally the home of Solomon Wiseman, the first settler here in 1808. The inn has accommodation and a restaurant that serves seafood, including local prawns and oysters. Phone (045) 66 4301.

On the River

There are **cruises** from Windsor on the Upper Hawkesbury every Sunday and Wednesday. Ring Windsor River Cruises on 621 4154. On the nearby Nepean River which flows into the Hawkesbury, a paddleboat cruise departs from Nepean Avenue jetty in Penrith. You'll watch the pretty gum tree covered slopes pass by to the music of Danny Boy on the player piano. Phone (047) 33 1274.

Or travel on your own by hiring a **houseboat** from Luxury Afloat Houseboats, located at the Hawkesbury Waters Leisure Park near Ebenezer, phone (045) 79 9253. Explore the historic Upper Hawkesbury or, if you choose, you can take the boat all the way down to Pittwater, about 100 kilometres from Windsor.

29   The Hawkesbury River and the Central Coast

The Hawkesbury River stretches all the way from the north west of Sydney down to the ocean at Broken Bay. The river is navigable as far as Windsor, about 100 kilometres from the mouth of the river. The English novelist Anthony Trollope gave this description to the river in the 1870s:

> ..the Hawkesbury has neither castles nor islands, nor has it bright, clear waters like the Rhine: but the headlands are higher, the bluffs are bolder, and the turns and manoeuvres which the waters have made for themselves are grander than those of the European River.

West, along the twisting river from Brooklyn, the houses are left behind and the river offers idyllic picnic grounds, good fishing and boating, and parts of the river are very popular water-skiing spots. Enjoy the river by boat, or following the river by car makes for a lovely drive. See the section above for a description of the historic towns on the Upper Hawkesbury. You can rent a cabin cruiser or houseboat and take a leisurely river trip. There are ferries and day cruises and one company offers overnight cruises.

Ferries from Brooklyn

Brooklyn is a small settlement near the mouth of the Hawkesbury River. The train from Sydney stops here. Hard as it is to imagine, this quiet place is named after Brooklyn, New York. The first rail bridge over the Hawkesbury was built by the company who built the Brooklyn Bridge and the little settlement adopted the name to honour the bridge builders.

The Hawkesbury River Ferries operate from Brooklyn and offer all sorts of possibilities for exploring the river. They run Australia's last riverboat mail run. The mail boat leaves Brooklyn every weekday morning and delivers mail, milk, bread and other essentials to the remote houses along the Hawkesbury. The boat leaves at 9.30 a.m. and returns at 1.15 p.m. The connecting train to Brooklyn station leaves Central Station at 8.15 a.m. On Wednesday, Thursday, and Friday there is an afternoon boat as well, departing at 1.30 p.m. Cost for the riverboat mail run is $14 for adults and $7 for children and pensioners. Lunch is extra, at $10 for adults and $5 for children. For bookings phone 455 1566.

The State Rail Authority organises a package day trip which takes in the mail run, a cruise around Broken Bay and a coach tour around the Central Coast. Tours depart from Central Station, Wednesdays and Fridays at 8 a.m. and return at 6 p.m. Cost is $49 for adults, $35 for children. For bookings phone 217 8812.

Hawkesbury River ferries run from Brooklyn to **Dangar Island** and **Wobby Beach** as well as from Brooklyn to **Patonga Beach**. All of these destinations offer wonderful day tripping. Many hikers head off to Wobby Beach to go walking in the Brisbane Water National Park. Dangar Island is a tiny island community and has a beach and picnic area. The ferry to Patonga will take you to long stretches of beach and sparkling waters. Ferries cost about $2 or $3 one way for adults and half price for pensioners and children. There is an extra charge for windsurfers, canoes, and bicycles. Ring 455 1599 for ferry times. There is also a ferry from the Palm Beach ferry wharf to Patonga, phone 919 5235 for ferry times.

Boat Hire

Another popular way to see the Hawkesbury is to hire your own houseboat, sailing yacht, or cabin cruiser. A boater's paradise, all of the Hawkesbury and Pittwater are waiting to be explored. For accomplished sailors, the open ocean is just beyond the estuary. Although boat rentals are expensive, your accommodation is at least taken care of. In fact it is not a bad alternative to staying in a Sydney hotel, as you can be off the river and into the city in about an hour. There are several different companies and here are a few examples of prices:

Halvorsen Boats (457 9011), Bobbin Head, Ku-ring-gai National Park. Cruisers, ranging from five to nine births, at motel accommodation rates, may soon be a thing of the past. Sadly, when going to press, it looked like the company would not be able to continue in business; still worth a call though, just in case!

Pittwater Yacht Charter (997 5344), Lovett Bay, Church Point. They hire fully equipped yachts of all sizes, costing from about $400 a weekend for a 24-footer (sleeps four) to a 38-footer (sleeps eight) for $1000. Rates are lower mid-week.

Gosford Area

Gosford is about 90 kilometres north of Sydney and has a population of 38,000. Many people commute to Sydney every day, most of them by train. Gosford is one of the fastest growing residential areas in the Sydney region, attracting those wanting to escape the crush of Sydney and to be near the water. The coast south and north of Gosford is lovely; many beaches such as **Avoca Beach** and **McMasters Beach** are popular with Sydneysiders. In the summer this whole area, including Terrigal and The Entrance, is a very popular holiday spot, with several caravan parks and motels. For more information on attractions and accommodation contact the Gosford City Tourist Association on 043 25 2835 or the Tuggerah Lakes Tourist Association on 043 32 9282.

Old Sydney Town

By road, take the expressway north of Sydney to the Gosford/Old Sydney Town turn-off, approximately one hour from Sydney. By public transport, take a train to Gosford station. A bus meets the train, leaving every half hour from 9.15 a.m. to 12.15 p.m., Wednesday to Friday. On weekends the first bus is at 9.45 a.m. There are also coach companies offering day tours to Old Sydney Town.

Old Sydney Town is a cross between an amusement park and a partial re-creation of Sydney as it was in its early days as a colony. Colonial town life is brought to you by townspeople dressed in period costume, going about their daily chores. Watch them unload a square rigger at port, or watch the chief constable try to bring a ruffian in line at the Magistrate's Court.

Traditional craftsmen making the necessities of daily life include a cooper making barrels, a blacksmith, a paper-maker, and a tinsmith. Street activities include Punch and Judy shows, minstrels, and wagon rides. The public buildings, shops, and thatched workers cottages are charming. Old Sydney Town is open during NSW public and school holidays, Wednesday to Sunday, from 10 a.m. to 5 p.m., at $12 per adult and $6 per child. Phone (043) 40 1104.

30   The Blue Mountains

The Blue Mountains are about 100 kilometres west of Sydney, which is about a two-hour drive. Take the Great Western Highway from the west of Sydney. Alternatively, there's a very good train service from Central Station to the mountains. Several tour companies and CityRail offer escorted tours to the mountains.

The Blue Mountains have been a popular place to visit since the 19th century:

> For those seeking an escape from the stifling heat, humidity and humanity of Sydney, a trip to the Blue Mountains is a must. Sought out by the tubercular and those after the contrast of the hill station with its breathtaking scenery and medicinal air, the well serviced guesthouses were all patronised. (Louise Johnson, *Gaslight Sydney*, 1984)

After World War II through to the early 1980s the Blue Mountains went through a decline. The virtues of the mountain air were substituted for sea air and the beach scene. But in the 1980s the mountains have seen a revival with many of the old hotels renovated. With good commuter trains and rising Sydney house prices, it has become increasingly attractive as a place to live. It continues to be popular with artists, actors, and writers and everyone who wants to escape Sydney. Many of the guesthouses and tea houses are run by former "ratracers".

The rugged mountain scenery and cool mountain air combined with a sense of the 19th-century resort lingering, gives the Blue Mountains a special atmosphere. The mood of the mountains seems to change dramatically depending on the weather. Often the Three Sisters can be invisible, shrouded in a damp mist or rain. But other days the sparkling sunshine sends you pounding down the hiking trail.

There's something here for everyone. Some like to go for a quiet weekend at a guesthouse where the furthest walk they have to make is to the dining room or to the lounge to curl up by the fire with a book. Others come for a camp-out in the gum forest by horseback. Most can't resist a stroll or a bushwalk along the escarpment overlooking spectacular sandstone cliff faces, gum trees, and waterfalls. Others come for the day hunting antiques and perhaps stopping at a tearoom of one of the grand old hotels. The tourist trips such as the cablecar and the exciting scenic railway down the gorge attract many visitors. The Blue Mountains can suit all budgets both for food and accommodation. Shop around a bit for both as some places can be very expensive. Reduced rates for accommodation during the week are common.

If you want to see some spectacular views without sharing it with coach loads of tourists there are several other dramatic

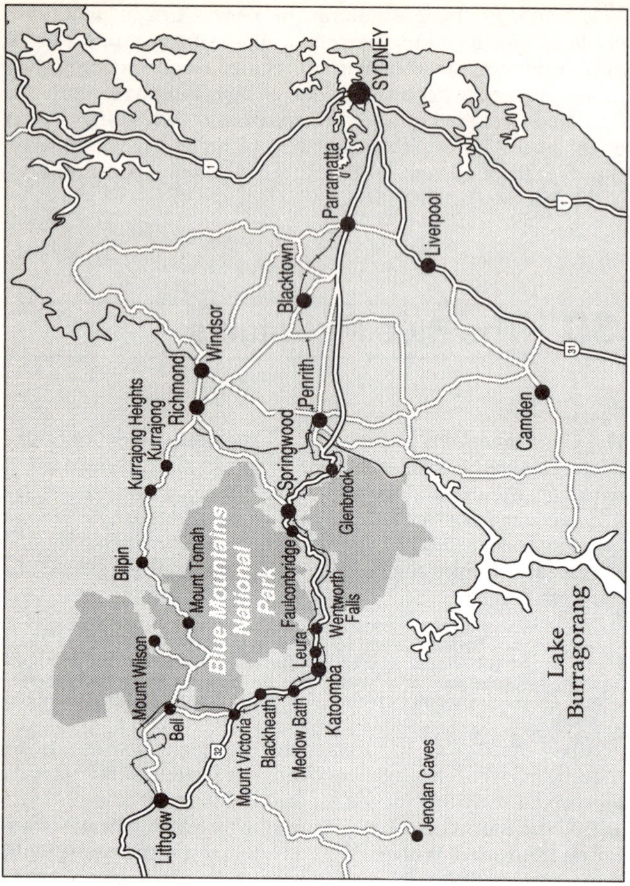

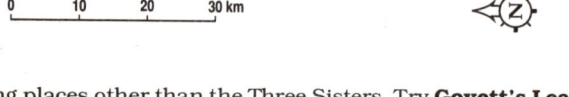

The Blue Mountains

0   10   20   30 km

viewing places other than the Three Sisters. Try **Govett's Leap Lookout** at the end of Govetts Leap Road at Blackheath, which has wonderful views over Grose Valley, complete with sheer rock faces, waterfalls, and towering gum trees. **Sublime Point Lookout** gives long distance views over Jamison Valley, and the **Megalong Road** at Blackheath gives an interesting view looking back at the ridge line. Superb Aboriginal carvings can be seen at King's Tableland, near Wentworth Falls.

Glenbrook-Faulconbridge

The National Parks and Wildlife Service has an information office at Glenbrook which is open weekends and many plesant bushwalks start from their centre. The **Norman Lindsay Gallery and Museum** is located at 128 Chapman Parade in Faulconbridge near Springwood. The famous artist Norman Lindsay lived and worked in this pretty sandstone house from 1912 until his death in 1969. The gardens and house now display his sculptures, paintings, and etchings. He also wrote many novels, including classic children's stories.

Wentworth Falls

The big attraction here is the falls which tumble down the dramatic rock face from great heights. Short and long walks can be made to view the rock face from Ingar Picnic Ground, on Queen Elizabeth Drive. By following the walk under the cliff you can weave your way down to beautifully cool waters for a splash.

A popular tearoom and restaurant in Wentworth Falls is the Elm Tree Tea Room for hearty meals such as steak and kidney pie, roast beef and yorkshire pudding, apple crumbles, etc. Open until late afternoon.

Leura

On the main street of the village of Leura is an interesting mix of craft and antique shops, restaurants and cafes. Leura is famous for its garden festival held every spring. **Everglades Garden** on Denison Street is a beautifully landscaped garden established in the 1930s, open daily from 9 a.m. Admission is $3 for adults, and children are free. Ring (047) 84 1938.

Katoomba

Katoomba is the main town of the Blue Mountains. The town itself doesn't do justice to its spectacular surrounds, although there are some lovely old buildings dotted around town. High-lights on main street are the Carrington Hotel and the Paragon Restaurant. The **Carrington**, built in 1880, has managed to retain some of the atmosphere of its high Victorian opulence. It has an art nouveau glassed-in verandah and a very elegant dining room. The **Paragon Restaurant** is an institution. Its facade as well as the lavishly ornamental chocolate trays decorating the window give you a hint of the restaurant's interior. It was established in 1916 and its interior is decorated in a fun art nouveau, art deco style. The cocktail room behind the restaurant is well worth a peak. Besides a good range of food from light snacks to full meals it has delicious chocolates.

Katoomba's main attractions are outside the town. **Echo Point**, famous viewing spot of the "Three Sisters" is about 2.5 kilometres from the main street of Katoomba. Sandstone pin-nacles rise over 300 metres from the valley floor. The "Three Sisters" were named by the Aborigines "Meenhi", "Wimlah",and "Gunnedoo" who, as the legend goes, were turned into stone for their misdeeds. The pinnacles are popular with rock climbers.

The **Scenic Railway** was built in the 1880s to carry miners into a coal mine. The sharp descent makes for an exciting ride down the tree-covered gorge. While waiting for the train you'll be entertained by crimson and eastern rosellas on the bird feeder. The **Scenic Skyway** cable car makes a dramatic cross-ing high above the valley. The railway and skyway are open daily from 9 a.m. to 5 p.m.

Blackheath

Blackheath is the next town west of Katoomba and is a popular base for bushwalking (see Part 6). On the way to Blackheath at Medlow Bath you'll be surprised by a giant hotel called the

Hydro Majestic Hotel, described as an Edwardian folly with a touch of art deco. Built in 1904, well before the age of environmental impact statements, it is right on the cliff's edge.

Jenolan Caves

About another 80 kilometres past Katoomba (180 kilometres from Sydney) are the remarkable Jenolan Caves. Wander through the 3 kilometres of limestone caverns and grottoes, all of course bearing fanciful names such as the "Indian Chamber in the Orient Cave", and the "Temple of Baal". Inspections are held between 9 a.m. to 5 p.m. and there are also night tours. Phone (063) 59 3304.

The only accommodation is Caves House which also has cabins (see section below). Golden West Tours operates a bus to Jenolan Caves from Katoomba and organises two and three day packages at Caves House. Ring (047) 82 1866.

The Bells Line of Road

The Bells Line of Road offers a pleasant drive back down the mountains and yet more wonderful views of the mountains. Among the sites along the way is **Mount Tomah Botanic Garden**, 12 kilometres west of Bilpin. This garden was established by the Royal Botanic Gardens for cool climate plants of the southern hemisphere. Open 10.30 a.m. to 6 p.m. from October to February and 10.30 a.m. to 4 p.m. March to September. Parking and admission is $5 per car. Phone (045) 67 2154. In season, Bilpin is a great place to buy crisp fresh apples.

Mount Wilson can be reached off Bells Line of Road. The cooler climate and its volcanic soils makes for outstanding plant variety. Many of the lovely private gardens are open in the spring with a gorgeous display of colour from the flowers and fruit blossoms. They are open again in the autumn to show off the beautifully coloured leaves of the deciduous trees. East of Mount Wilson the best of the mini rainforests can be viewed at the **Cathedral of Ferns**. The short path is shaded by towering tree ferns and there is a pleasant picnic area here too.

In between the town of Bell and Lithgow on the Bells Line of Road is the delightful **Zig Zag Railway** built in 1869 as part of the main western railway line. It was closed in 1910 and is now a tourist train. The steam train zig zags its way down the Lithgow Valley through tunnels, over viaducts and along steep ledges making for a thrilling ride. Operates weekends only, phone (063) 51 4826.

Information and Accommodation

The Blue Mountains Explorer bus running on weekends in the Katoomba-Leura area makes sightseeing without a car much easier, phone (047) 82 1866. The tourist information centre is located at Echo Point, phone (047) 39 6266. The complimentary "Blue Mountains Experience" is a helpful visitors' guide listing bushwalks, restaurants, accommodation, and activities, and is available at the tourist information centre. There are countless bushwalks, ranging from an easy stroll to extremely challenging endeavours. Contact the tourist information centre above, or National Parks and Wildlife at the Heritage Centre, Govett's

Leap Road, Blackheath, on (047) 87 8877. For overnight camp-outs by **horseback** or hourly riding, ring the Packsaddlers on (047) 87 9150.

The CityRail Blue Mountains day tour is good value at $29 for adults and $12 for children. A sightseeing coach meets the train at Katoomba Station. Departure from Central Station is at 8.19 a.m., 9.20 a.m., and 10.50 a.m., Monday to Friday. Phone 29 7614. Several coach companies offer one-day tours also, such as Ansett Pioneer, AAT Kings Coaches and Deanes Clippers Tours.

There are dozens of places to stay in the mountains—from youth hostels to an international hotel. Some of the gues-thouses are open on weekends only. Here are a few suggestions:

The *Katoomba Youth Hostel* (047-82 1416) has two locations, one on Wellington Road and one on Waratah Street. About $10 per person per night.

Glenella (047-87 8352), 56 Govetts Leap Road, Blackheath, has comfortable rooms and a dining room overlooking the lovely gardens. It is renowned for its gourmet cuisine and is often booked up months in advance for weekends. Overnight stays are from $185 per double.

Little Company Guesthouse (047-82 4023), 2 Eastview Avenue, Leura. If you're looking for a game of croquet or lawn bowls this guesthouse offers both. Bed, breakfast and dinner for $95 per person a night.

Jenolan Caves Resort (063-59 3304) or toll free (008-02 7927). This charming hotel at the caves is very popular, so book well ahead in the tourist season. They also have cabins.

Holiday Cabins set in the woods are a good way to experience the mountains, and they are economical. There is *Werriberri Lodge* (047-87 9127) on Megalong Road, Blackheath, or *Jemby-Rinjah Lodge* (047-87 7622) on Evans Lookout Road, Blackheath. They are fully self-contained with open fires or wood stoves and start at about $50 per night.

31   The Hunter Valley

Cessnock, the main centre closest to the wine area is about 180 kilometres from Sydney. By car you can take the Princes High-way (about 2.5 hours) north to Newcastle then through Hexham to Cessnock. An alternative, more scenic route is to turn off the highway at Peats Ridge and take the Wollombi Road. It is not possible to see the wineries by train, but CityRail and coach com-panies offer tours. Batterhams Busline has regular buses to Cessnock (phone 27 2672), and there are also local buses from Newcastle and Maitland railway stations.

The Hunter Valley is famous for its wineries. It is the oldest wine producing area in Australia, dating from the 1830s. Over forty wineries, most of them around Pokolbin, share the Valley. The wine-tasting circuit is a popular activity with Sydneysiders. Small wonder, the wine is free and the scenery lovely. Try a weekday visit as you're likely to have a lot more elbow room and no trouble finding accommodation.

Besides the wineries there are other attractions in the Hunter, including hot air ballooning, hiking at Barrington Tops, exploring heritage sites, and horseback riding and golf.

Touring the Wineries

The serious wine seeker should take one of the Australian wine guidebooks along (see Further Reading). Most of the wineries are open from 9 a.m. to 5 p.m., Monday to Saturday, and they usually open from about noon to 5 p.m. on Sundays. Many of the wineries also have fine restaurants as well as picnic areas.

Most of the wineries are located between the towns of Pokolbin, Branxton and Broke. At the Pokolbin end of the wine trail, **McWilliams** at Mount Pleasant has a lovely site in the valley against the hills. Guided tours are at 10 a.m., 11 a.m. and 12 noon, Monday to Saturday. The old winery of **Lindeman's** is a popular spot for its atmosphere, Hunter burgundy and the semillons. The **Hungerford Hill Wine Village** is a sort of one-stop shopping spot with accommodation, restaurants, shops and lots of play equipment to occupy children. Their chardonnays have a particularly good reputation. The **Rothbury Estate** has gained a reputation for its semillon and shiraz but is moving more into chardonnays. The biggest wine producer in the Hunter Valley is **Wyndham's Estate**. On the Broke Road, **Tyrell's** is an old name in the Hunter, being one of the original wine producing families. The old winery is a delight with its wooden casks and hand presses. A bit off the beaten path at Broke is Saxonvale where decent wine at good prices is on offer.

Other Sights in the Hunter Area

Maitland is an historic town well worth a browse. It has a long heritage associated with coal mining and agricultural activities. A fascinating gaol from the 1840s on King Street is now a museum displaying prison, police, and bushranger gear. Opening hours, Monday to Friday, 10 a.m. to 4 p.m., and 10 a.m. to 5 p.m. on weekends. Many other interesting buildings in the town date from early colonial times.

Newcastle may not look very impressive as one skirts around it, en route north or to the wine country, but it is much more than an industrial city. The second largest city in New South Wales, it now has a population of about 260,000. The pedestrian mall in the centre is the city's focus, with shops, restaurants, and fine old buildings. A severe earthquake in late 1989 damaged many of Newcastle's old buildings, and the City Council unfortunately ordered a number of them to be demolished. Excellent beaches, lovely picnic areas, as well as **Fort Scratchley** and its **Military and Maritime Museum** are among the many attractions.

The **Gloucester-Barrington Tops** area is a remote corner of the Hunter. Gloucester has a lovely setting in the hills. Barrington Tops National Park stands in sharp contrast to the warm climate and gently rolling hills of the Hunter Valley. At about 1500 metres above sea level it has alpine vegetation, such as the twisted snow gums and snow grass plains reminiscent of the Snowy Mountains. Dingoes, so they say, can be heard most nights in this rugged country. Further down the slopes lush sub-tropical rainforest replaces the hardy snow gums. No wonder this is a popular place with bushwalkers. For those who prefer the comfort of a bed and a chair on a verandah to a tent, the *Barrington Tops Guest House* is situated on the plateau. Several walks start from the guesthouse and horses can be hired for $5 an hour (see Accommodation below).

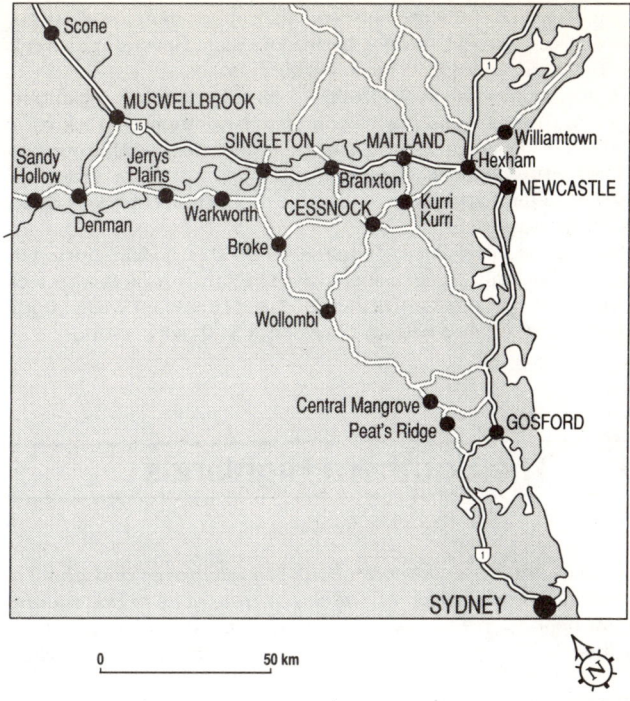

The wide open spaces and lovely scenery make for great ballooning country. **Hot air balloons** depart from Hungerford Hill Wine Village (049-38 1955) and the Pokolobin Trading Post (049-98 7633). The cost is around $100 per person.

Information and Accommodation

There is a **tourist information centre** at Cessnock, corner of Mt View and Wollombi Roads (049 90 4477), open every day 9 a.m. to 5 p.m. Newcastle also has a very helpful tourist information centre on Queen's Wharf in Newcastle, ring (049) 29 9299.

Taking a **tour** is a good way to follow the wine trail if you don't want to worry about drink-driving. City Rail offers a day tour to the Hunter Valley, departing around 7.30 a.m. from Central Station and returning about 7.30 p.m. The fare includes all rail and bus transport and lunch, at $69 per adult and $42 for children. Tours are on Tuesday, Thursday, Saturday, and Sunday. Phone 217 8812. Coach companies also offer tours.

The accommodation among the wineries is generally more expensive than in nearby Cessnock. There are several good, reasonable places to stay in Newcastle, which is only about a 45-minute drive from the wine area.

The *Bellbird Hotel* (049-90 1849) on Wollombi Road is a popular place. It is a country pub with wide verandahs. Rooms are about $50 a double. Book ahead for weekends.

Black Opal Hotel (049-90 1070), 216 Vincent Street, Cessnock, is pub-style accommodation, and the *Youth Hostel (YHA)* is also located in this hotel. Hostel accommodation is $9 a night.

Pokolbin Village (049-98 7670), 188 Broke Road, Pokolbin.

In the heart of the wine country, this offers motel style units and self-contained 2- and 3-bedroom villas. Special mid-week rates start at about $70 for a double room.

Peppers Guest House (049-98 7596), Ekerts Road, Pokolbin, is one of those special places you go to be pampered. Pick your method, there is a masseur, sauna, indoor pool, and library as well as tennis courts. The Australian colonial decor gives it a cosy yet elegant atmosphere. Double rooms start at about $120 and meals are extra.

Barrington Tops Guest House (049-95 3212), Salisbury, via Dungog. The only place to stay at Barrington Tops, it is a perfect place for a bush retreat. Dinner, bed and breakfast costs about $45 per person, and bed and breakfast $30 per person.

32   The Southern Highlands

The Southern Highlands are about 115 kilometres and about a 2-hour drive south-west of Sydney. A train goes to Bowral and Bundanoon. Berrima Coaches run an express bus service from Sydney to Picton, Mittagong, and Moss Vale; phone 281 2233 for timetable.

The Southern Highlands is another one of Sydney's playgrounds. Not as dramatic as the Blue Mountains, Morton National Park to the south nevertheless has some stunning scenery. The rural roots of the Highlands are strong, and many of the country's top cattle and horse studs are here. The lush green fields and pretty towns are reminiscent of Tasmania, both in landscape and heritage.

First settled in the 1820s, its development was considerably boosted when the railway came through in the 1860s. Many wealthy Sydneysiders have built retreats here and fine estates dot the countryside. The towns in the Highlands are small, some on the touristy side but not unpleasantly so. Tourist facilities are very good, with golf courses, swimming pools, tennis, hiking, horseback riding, fine restaurants and hotels. Two roads lead to the coast from this area providing a lovely way to see some of the coast and inland scenery on the return trip to Sydney. These routes—Kangaroo Valley Road and the Jamberoo Road—are also described in this section. The Highlands towns described below have been selected to give a flavour of the area and neither include all the towns nor all the sights.

Bowral

Bowral is the commercial centre of the Highlands with a population of 8500. It can be reached by train from Sydney and there is accommodation available within walking distance of the station. Founded in the 1860s many of its impressive buildings date from the boom times of the 1880s. The main street is a bustling retail strip, with a variety of craft and antique shops, restaurants, food and clothing shops. In addition to the fine old buildings on the main street, there is a lovely block of municipal buildings on Bendooley Street, one block east of the main street. A good place for a picnic, Corbett Gardens, is also

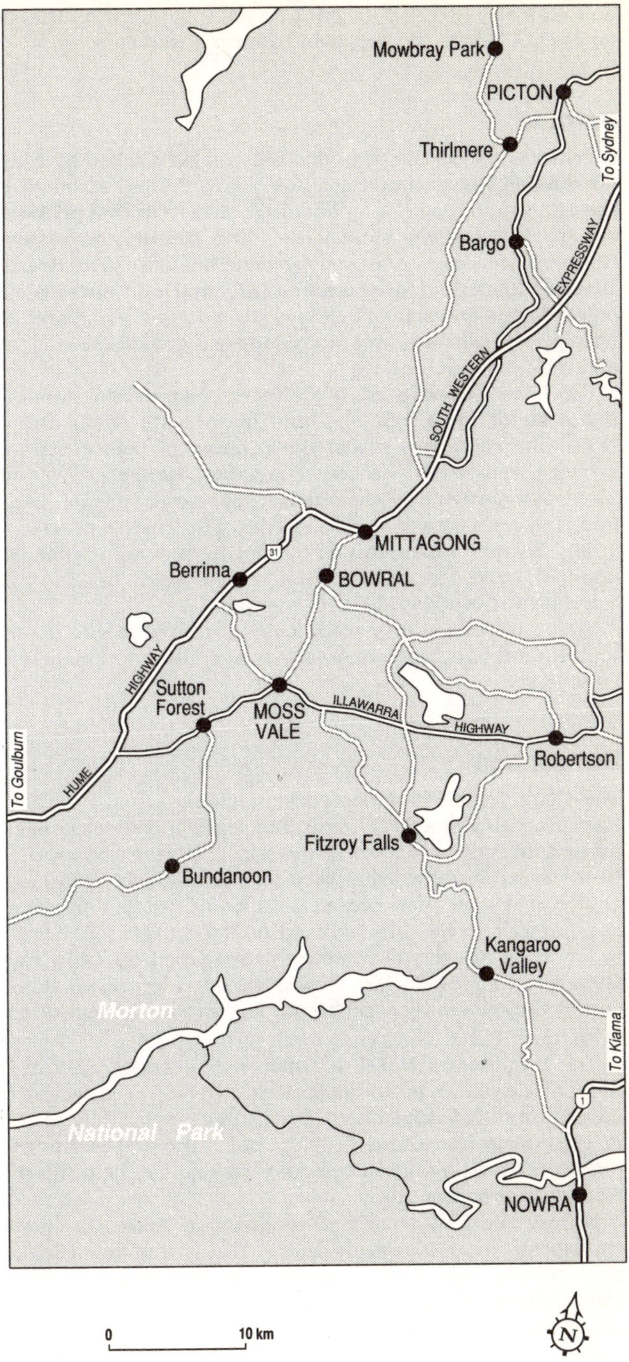

Mowbray Park

PICTON

Thirlmere

To Sydney

Bargo

SOUTH WESTERN   EXPRESSWAY

31

MITTAGONG

Berrima

BOWRAL

HIGHWAY

Sutton
Forest

MOSS
VALE

ILLAWARRA   HIGHWAY

Robertson

To Goulburn

HUME

Fitzroy Falls

Bundanoon

Kangaroo
Valley

Morton

To Kiama

National   Park

1

NOWRA

0   10 km

on Bendooley Street. A recent and not insignificant attraction for cricket fans is the new **Don Bradman museum**.

Berrima

Berrima is not on the train line but can be reached by getting off at Moss Vale and taking a bus. Berrima was established in the 1830s and many of its buildings date from that period. An important travellers' stop in the 19th century, as many as thirteen inns once competed for their business. The **Berrima District Historical Museum and Information Centre** is open on weekends and school holidays (10 a.m. to 4 p.m.), and they have pamphlets for a very interesting self-guided walking tour, about 2.5 kilometres long.

On Jellore Street there is a cluster of sandstone buildings. **Brian McMahon's Pub** was built here in the 1840s and the worn window ledges reveal the number of beers that were served through the window. The **Court House** on Wiltshire Street is a particularly fine building, open every day for inspection. To get an idea of a typical home of the middle class in the 1830s, **Harper's mansion** on the northern outskirts has been restored by the National Trust and is open on Saturday, Sunday, and Monday, 11 a.m. to 4 p.m.

Plenty of restaurants around town will tempt you on your tour, the fare ranging from Aussie damper to exceptional French cuisine.

Bundanoon

The character of Bundanoon is very different from Bowral and Berrima with just a few restaurants, cafes and shops clustered around the railway station. At the edge of Morton National Park, Bundanoon is a favourite place for bushwalkers. You have a choice of several short pleasant walks as well as long ones. A popular way to reach the walks without a car is to rent a bicycle in town and ride the 3 kilometres or so to the trail. **Glow Worm Glen** has a large colony of glow worms in a fern-lined glen. Of course they glow only at night, but it is an easy 20-minute walk from town. Don't forget your torch or flashlight.

The **Bundanoon Hotel**, a Tudor style surprise right in the heart of the village, is old-fashioned and a bit frayed around the edges, but full of atmosphere. It is a large hotel with good hearty breakfasts and dinners. A special attraction is the roast dinner followed by an afternoon of jazz music, held the third Sunday of every month.

During the first or second weekend in April, Bundanoon transforms into Brigadoon and is the site of the Highland games. Hundreds of bagpipers from clans and clubs all over New South Wales descend upon the village and parade through the streets. The games include hurling the haggis, and the caber and egg toss, of course.

Information and Accommodation

There is plenty of accommodation in and around the Highlands to suit all tastes and budgets.

Mowbray Park (046 809 243), near Picton, is the place for a **horseriding** holiday or a quiet rural retreat. The historic farm buildings are set in beautiful bush and countryside. Facilities include a swimming pool and tennis courts. Bed and breakfast is from $27 a day for adults and $15 for children, horseriding extra.

Milton Park (048 61 1522), Horderns Road, Bowral. Referred to as a six-star country hotel, this is ranked among the world's best. Once the rural retreat of the retail giant Anthony Hordern, the gardens are magnificent, and all luxury facilities are here, including swimming pool, tennis, golf, and even croquet. At $120 to $300 for a double room and continental breakfast it may be the place for your splurge. The dining room serves luncheon and dinner and is not restricted to hotel guests.

Bundanoon Hotel (048 83 6005), Erith Street, Bundanoon. This is a large hotel from the 1920s with a swimming pool and tennis court. Unfortunately the rooms do not have the charm and cosiness of the billiard and lounge rooms and need updating, but they are adequate. Rates for bed and breakfast start at around $45 per person, and for dinner, bed and breakfast at about $65.

The *Youth Hostel (YHA)* (048 83 6010) is less than a 5-minute walk from the train station on Railway Avenue. Cost is about $8 a night per person.

Kangaroo Valley

Kangaroo Valley is a scenic valley running between Nowra near the south coast and the Southern Highlands, about 60 kilometres from Nowra. The round trip from Sydney to the Highlands, through Kangaroo Valley, then up the coast back to Sydney, can be made in a day. It is better to take two days over the 340 kilometres, to leave time for browsing or long lunches. The Kangaroo Valley road weaves its way down the mountain side through lush rain forest.

Fitzroy Falls is a particularly lovely spot with waterfalls tumbling down through the rainforest and tree ferns galore. There are short and very pleasant bushwalks here.

There are two routes to the coast from Robertson. The southern route, Jamberoo Road, is the more interesting. Driving is slow because of the winding road and steep terrain. The road is sealed except for an 8-kilometre stretch along the Jamberoo Pass and it is not advisable to take this route in wet weather. The scenery along this road is stunning and you'll pass by Carrington Falls and Minnamurra Falls.

33 The South Coast

Follow the Princes Highway south out of Sydney and along the coast. Wollongong is 82 kilometres from Sydney, and Kiama is 122 kilometres away. A nice alternative to car travel is the train, as it offers some lovely views of the coast on the way to Kiama. Coach tours are also available.

The beautiful **Royal National Park** lies to the south of Sydney (see Part 6), and beyond the park the scenery along the coastline changes from golden beaches to rugged rocks. The viewing point at **Bald Hill Lookout** near Stanwell Park will give you a great view of the spectacular coastline. This is a favourite place for **hang gliding**.

Wollongong and Beyond

Wollongong is an industrial town, the site of steelworks and a port. The third-largest city in New South Wales, its population is about 210,000. The city has many attractions including museums, botanic gardens, and several lovely beaches. The revitalised shopping area and pedestrian mall offer good shopping, restaurants, and cafes. The Italian Club in Burelli Street has excellent food and good strong coffee.

Kiama is an interesting town perched right on the ocean. You can take a walk out to the famous **blowhole** on the point just beyond the town. The ferocious sea forces water up through a hole in the rock giving the effect of an enormous whale spout. There are a few craft shops, antique shops, restaurants and cafes as well as a museum at the Boat Harbour. There is also a nice beach and plenty of picnic spots around town.

Jamberoo is inland from Kiama and has a few interesting attractions. The **Minnamurra Falls Reserve** is a sub-tropical rainforest on the Jamberoo Pass Road. It's a short and pleasant walk to the Falls which plunge into a deep rock gorge, and the picnic grounds are very pleasant. The **Barrens Grounds Nature Reserve** is also on the Jamberoo Pass Road and is a favourite place for bushwalks and birdwatching. The hanging swamp plateau is almost completely surrounded by sheer cliffs. The **Jamberoo Grass Ski Park** is north of Jamberoo on the Albion Park Road. Open on weekends, it has grass skiing, a mountain bobsled course, a swimming hole with diving off the rocks, and a chairlift. Open weekends from 10 a.m. to 5 p.m., ring (042) 36 0114.

The **Seven Mile Beach National Park** is about 140 kilometres south of Sydney and 17 kilometres east of Nowra. You can get great views of this stretch of beach from Black Head at Gerora, south of Gerringong. High sand dunes line the beach and a pleasant, shady picnic area in the woods behind the beach has toilets, barbecues, and tables.

Information and Accommodation

There are several motels around Wollongong, Kiama and Nowra. The *Kiama Inn* (042-321 166), 50 Teralong Street, offers bed and breakfast for about $40 for a double room. *Gundarimba Guest House* (042-36 0228), 45 Allowrie Street, is a stately 1920s mansion set in lovely gardens, offering first-class accommodation and fine food. Cost per person is about $125 for dinner, bed and breakfast.

34   The Snowy Mountains

The Snowy Mountains are about 450 kilometres south-west of Sydney, which is a 7- to 8-hour drive. Alternatively, you can fly to Cooma where buses meet the flights. A good coach service links Sydney to the ski resorts and there are also package bus trips from Sydney.

Many North American visitors coming in the winter months are curious to try the ski slopes "down under". And for Australians visiting the east coast it is an easy and exciting side trip. The ski season normally runs from June to early October. Although the Snowy ski resorts are smaller than most of the North American and European Alps resorts, they offer very good skiing conditions for both beginners and experts. The *apres ski* is not bad either and the centre of night-life is in Thredbo. All the major resorts provide ski rentals, clothing rentals, lessons, and child care. The average price of a day ticket at the major resorts is $40. The 5-day passes and 5-day lift/lesson packages offer good discounts.

The Snowies are a popular summertime destination too. The wildflowers in the mountain pastures are a wonderful sight. Mount Kosciusko, Australia's highest mountain at 2260 metres, is an easy hike. Some of the chairlifts are open in the off-season, offering views of the mountains.

There's a large choice in hiking trials from easy to difficult in the national park.

Ski Resorts

Perisher/Smiggins is the largest resort, and is at the highest altitude of all resorts. It has 30 lifts and the vertical rise is 375 metres. It has an enormous ski school with over 100 instructors. Accommodation is available at the slopes.

Thredbo has the longest ski runs, the highest vertical rise at 670 metres and 15 lifts. The village of Thredbo has a distinct alpine feel with the Swiss chalets and other village buildings hugging the hillside.

Mt Blue Cow is a day resort without accommodation. Get there by the skitube (ski train which costs $10) from Bullock Flat.

There are three other small resorts, **Guthega**, **Charlotte Pass**, and **Mt Selwyn**. Some of the smaller resorts offer good discounts for beginners and families.

Good skiing can be had further afield across the border in Victoria. The main resorts there are Falls Creek, Mt Hotham and Mt Buller.

Information and Accommodation

Accommodation right at the slopes of Thredbo and Perisher/Smiggins is expensive and often booked up, so shop around and book in advance if possible. Self-contained apartments are available and can be a better buy than motels and hotels for families or a group of friends. The youth hostel, Jack Adams Patch, is at Thredbo, phone (064) 57 6376.

Jindabyne is a good bet for more affordable accommodation.

It is about a 30-minute drive to Thredbo and Perisher from here, and there is a good bus service. Jindabyne is the largest commercial centre in the mountains with supermarkets, a good variety of other shops and services, and an ice-skating rink.

Most travel agents or the Travel Centre of New South Wales in Sydney (231 4444) will help with your travel arrangements. Also check the newspapers for package deals. Toll free telephone calls can be made to the Jindabyne Reservation Centre on (008) 026 356. For a recorded message on ski conditions and accommodation, ring 11 539.

PART VIII
Things to Do

The year in Sydney

Museums, art and craft galleries

Entertainment

Restaurants

Shopping

Sports and outdoor activities

Spectator sports

The year in Sydney

This section provides a quick look at the main festivals, fairs, and special annual events. Spectator sports and events are listed separately at the end of this section.

January

Festival of Sydney (the entire month of January) January is school holiday time and this month-long festival stages events for all ages. There's plenty of free entertainment around the parks and Darling Harbour, like rock concerts, theatre, and folk dancing.

The **Manly Ironman Gold**, usually the second or third week in January, is an Australian cultural event not to be missed. Water heroes of back of breakfast cereal-box proportions battle it out in the sea and the sand. Held at Manly Beach, South Steyne, ring 597 5588 or 977 1088.

The **Great Ferry Boat Race** (26 January) with the start and finish line at the Harbour Bridge is a fantastic spectacle. The harbour ferries decorated with balloons and banners race around the harbour. There are plenty of other nautical events, including the spectacular 12-metre yacht races. The free **Opera in the Park** with world famous opera singers draws crowds in excess of 100,000. The finale of the **Symphony under the Stars** concert usually sends the audience in a tizz with the mighty 1812 Overture, cannons, fireworks and all. Free programs of the festival available all around the city. Ring 281 6000.

Australia Day (26 January) This holiday commemorates the founding of European settlement in Australia on 26 January 1788. The focus is on the harbour with The Great Ferry Race (see above) plus lots of other activities including an impressive fireworks display over the Harbour Bridge.

Chinese New Year (usually around the end of January to early February) Chinatown becomes even more lively with a parade and Lion Dances in Dixon Street. The Chinese restaurants are jam packed on this day so book ahead.

March

Gay Mardi Gras (usually early March) The Gay Mardi Gras is a wonderful parade which draws crowds of over 200,000 people. In atmosphere it is the closest Sydney gets to Rio's Carnival. It is usually held on the first Saturday of March and starts around 8 p.m. It travels down Oxford Street and ends at the Showgrounds (Hyde Park), where the partying continues all through the night.

April

The Royal Easter Show (Easter time, beginning one week before Good Friday and ending Easter Tuesday) The Show is a Sydney institution and has been held annually for over one and a half centuries. This is the time of year when the country folk come to the big smoke. Attractive displays of farm produce fill the halls, and all are out in pursuit of that illustrious blue ribbon, entering everything from jungle fowl to lamingtons (a special Aussie sponge cake rolled in chocolate and coconut). There are rodeos, sideshows and evening entertainment. Ring 331 9111 for more information.

Anzac Day (25 April) This is a day of remembrance for those who served in the wars and is a public holiday. It originated as a commemorative event for the soldiers of the Australian and New Zealand Army Corps (ANZAC) who died in the Gallipoli Campaign against the Turks in 1915. It is an important day for Australians, with thousands participating in a march through the city centre and ending at the War Memorial in Hyde Park.

June

Sydney Film Festival (two weeks beginning Queen's birthday long weekend) This international film festival held at the magnificent State Theatre draws large crowds and often keeps many Sydneysiders away from work for days. The festival director scours the world's festivals to bring the best of the international films as well as the best in Australian feature films, shorts and documentaries. Ring 660 3844.

National Folkloric Festival (usually first or second weekend in June) A magnificent display of international folkdancing and singing at the Opera House with a cast of over 1500 representing 50 cultures. The festival begins with a parade around Circular Quay to the Opera House. Ring 250 7111.

Biennale (usually held half way through every even-numbered year and goes on for a few weeks) Modelled after the Biennale of Venice, this is an international art festival featuring paintings, sculpture, photography as well as music and performing arts from Australia and around the world. Works are exhibited at the Art Gallery of New South Wales and Pier 2/3 on Hickson Street, Walsh Bay.

Manly Food and Wine Festival Attracts large crowds to sample wines and food from dozens of stalls.

August

City to Surf Race (second Sunday in August) This 14 kilometre fun run begins near Hyde Park and finishes at Bondi Beach. It's quite a spectacle because of the sheer numbers, attracting close to 40,000 runners. Some of the runners are world class, others come for the fun of it and can be seen pushing prams, dressed in gorilla suits or in black tie with a tray of drinks. The dogs are always popular and receive great audience cheers as they cross the finish line.

September

Carnival (second week in September) This is a cultural festival running for two weeks throughout New South Wales. Ethnic groups organise events like dancing, theatre, and films. The blessing of the fishing fleet is a popular event at Darling Harbour and a big woolshed dance is held at Flemington markets. The Festival of the Winds at Bondi is a kite-flyer's paradise—you can buy one there and join in or sit back and watch the colour.

October

Manly Jazz Festival (Labour Day long weekend, usually first weekend in October) The Corso comes alive to the sweet sounds of jazz with free outdoor concerts put on by international performers. The Manly ferries even have jazz bands during the festival. Phone 977 1088.

Kings Cross Carnival, usually on the second weekend in October, is a street festival with concerts, stalls, food, etc.

December

Sydney to Hobart Race (26 December) The day after Christmas thousands of Sydneysiders watch the start of this great ocean sailing race from vantage points around the harbour or from their boats. If you book early you can escort the fleet out to sea on a ferry, phone 27 4738.

Museums, art and craft galleries

The museum scene in Sydney has changed considerably in the last few years. The old museums have shed their stuffy image, gone through major renovations and have added lots of exciting exhibits. Even the once very sombre State Library welcomes the public "to visit our magnificent Macquarie Street complex for a dynamic experience" and "access unlimited information in comfortable surroundings". Many new museums have also sprouted up, most of them small but with very interesting collections.

The Australian Museum (339 8111) corner College Street and William Street, Open Tuesday to Sunday, 10 a.m. to 5 p.m., Monday, 12 noon to 5 p.m. Admission free. This natural history museum is no longer a dusty storage place for stuffed animals. There is an excellent display of Aboriginal artefacts with a reconstructed archaelogical site where stone tools, food remains, and campfires were studied. In the Papua New Guinea room, the day to day life of a Sepik River village is displayed, along with the facade of a beautifully ornate ceremonial hut. In the "hands-on" discovery room you can reconstruct skeletons or put together jigsaw fossils. The bird gallery has a large collection of colourful Australian birds complete with sound recordings of bird calls. Another fascinating exhibit is "Dreamtime to Dust" showing how the Australian environment has changed over the last 200,000 years, highlighting the impact of the Aborigines and the Europeans.

The Powerhouse Museum (217 0111), corner of Macarthur Street and Harris Street, Ultimo. Open daily 10 a.m. to 5 p.m. Admission free. Ring the Powerhouse Hotline for what's on: 11600. This is Australia's largest museum and it really does convey a feeling of wonder and excitement. There are vast collections for social history, science and technology, and the decorative arts. Proving that *learnum est funum* the museum has the atmosphere of a fun palace and is very popular with children. Here are some highlights:

"Everyday Life in Australia" consists of five exhibits about the lives and times of ordinary people. One exhibit is on the Aborigines of La Perouse in Sydney. Another is on Australian women's work in the home called "...never done". There's an 1880s bush hut which shows a few problems with old kitchen designs. Learn about Australian drinking habits in "Brewing and the Pubs".

"Experimentations" is a science-in-action exhibit with plenty of hands-on experiments.

"Transport": a Catalina flying boat hung from the ceiling is the largest suspended exhibit hanging in any museum in the world. You'll also see other transport exhibits, like trams, motorcycles, and trains.

The museum shop is full of unusual objects such as beautiful designer stationery, glass works, jewellery, experimental kits, and books. There is also a coffee shop and an outdoor kiosk on the courtyard.

Hyde Park Barracks Museum (217 0111) Queen's Square, Macquarie Street, open 10 a.m. to 5 p.m., except Tuesday, 12 noon to 5 p.m.. Admission free. This is a museum spanning the history of Australia, displaying its everyday life and culture.

Sydney Observatory (241 2478) Observatory Hill. Open 2 p.m. to 5 p.m. weekdays and 10 a.m. to 5 p.m. weekends. Admission free. The observatory has a "hands-on astronomy" display, where visitors can play and experiment with gravity, lenses, light, elliptical orbits, and so on. They also have night viewings at 8 p.m. but bookings are required.

Mint Museum (217 0111) Macquarie Street. Open 10 a.m. to 5 p.m., except Wednesday, 12 noon to 5 p.m. Admission free. Museum of decorative arts, coins, and stamps.

The Earth Exchange (251 2422) 36 George Street. Open Monday to Friday, 9.30 a.m. to 4 p.m., Saturday, 1 p.m. to 4 p.m., Sunday, 11 a.m. to 4 p.m. Hours may change, check before visiting. Formerly called the Geological and Mining Museum, it has been completely refurbished and expanded. The new museum has a massive collection of over 300,000 items and has lots of exiting computer games and working exhibitions. One display takes you across a wooden walkway

Maritime Museum

into the landscape of a Hawaiian volcano at night followed by a walkway that twists and turns to the movement of an earthquake. The shop has specimens for sale as well as lapidary items. The restaurant on the top floor has spectacular views over the harbour and Opera House.

Australian National Maritime Museum (552 7777) Darling Harbour. Open daily 10 a.m. to 6 p.m. This exciting new museum is expected to open in 1991. Situated right beside working docklands, it is in an ideal location. It is much more a museum of Australian maritime social history than a mere collection of nautical bits and pieces.

Australia II, the successful America's Cup challenger, is in full sail inside the museum. The museum's collection extends out into the wharves in front where many boats are moored. The full scale replica of the *Endeavour*, a pearling vessel from Broome, and a Navy Destroyer are among the many boats on the water. There's a "hands-on" Discovery Room for children, a wharfside restaurant, and a museum shop.

La Perouse Museum (661 2765) End of Anzac Parade, La Perouse. Open daily 10 a.m. to 4.30 p.m. Adults $2, children and concessions, $1. This museum traces the voyage of La Perouse, the French explorer who landed in Botany Bay just a week after Captain Phillip. Through maps, artefacts, and engravings the museum tells the intriguing story of the voyage right from the expedition planning stages to the discovery of La Perouse's shipwreck.

Captain Cook's Landing Place Museum (668 9923) Kurnell, Botany Bay. Open daily, 10.30 a.m. to 4.30 p.m. weekdays and 10.30 a.m. to 5 p.m. on weekends. Admission fee per car is $3.50. This fascinating museum shows Cook's voyage and achievements, examples of specimens collected by the botanist Joseph Banks, and an explanation of the Transit of Venus.

Maritime Services Board Harbour Museum (272 2733) Goat Island. The museum, tracing 150 years of maritime history, is located in the former "Officers Quarters" built in 1838. The island was used as an explosives store until the turn of the century. See "The Harbour Islands", Part 5 for times of guided tours.

Quarantine Station (977 6229) North Head near Manly. Guided tours, bookings essential. Admission $2 adults, $1 children. The quarantine station was used to protect the country from infectious diseases such as scarlet fever and influenza. It is now being restored by the National Parks and Wildlife Service. The site has Aboriginal rock carvings and historic buildings.

The Australian Museum of Childhood (332 1988), Juniper Hall, 248-250 Oxford Street. Open daily 10 a.m. to 4 p.m. Admission $3 for adults, $1.50 concessions. A National Trust property, this building holds a collection of 19th and 20th century children's books, toys and illustrations. There's a cosy coffee shop and book and gift shop.

Children's Museum (897 1414) corner of Pitt and Walpole Street, Merrylands. Open Tuesday to Friday and Sunday from 10 a.m. to 4 p.m. Admission $3 adults, $2 children. This is a hands-on museum with lots to interest children, including a science area, a hospital room complete with skeletons, a dental surgery, and a historical room.

Medical and Nursing Museum Marsden Street and George Street, Parramatta. Open Sunday, 10 a.m. to 4 p.m.

Regimental Museum Linden House, Lancer Barracks,

Smith Street near Station and Macquarie Streets, Parramatta. Open Sundays 11 a.m. to 4 p.m. Admission $1 adults, 50 cents children. The Royal New South Wales Lancers display the history of their regiment, which was established in 1885.

Victoria Barracks Military Museum (339 3543) Oxford Street, Paddington. Changing of the Guard every Tuesday at 10 a.m. The museum opens after the ceremony until 12.30 p.m. and is also open on the first Sunday of each month, 1.30 p.m. to 4.30 p.m. The Barracks are closed to the public from the middle of December till the end of January. Admission free. The museum has a collection of military memorabilia such as uniforms and historic weapons.

Justice and Police Museum (223 5090) corner Alfred and Phillip Streets, near Circular Quay. In the old Water Police Station and Court House, it houses a collection of displays about the history of the NSW police and justice system. The courtroom holds mock trials.

The Museum of Fire (047-31 3000) Castlereagh Road, Penrith. Open Monday to Friday, 10 a.m. to 4 p.m. and Saturday and Sundays, 10 a.m. to 5 p.m. It describes itself as a "museum and education centre exploring the drama and danger of fire". It has steam train rides on Sundays.

Thirlmere Railway Museum (046-81 8001) Barbour Road, Thirlmere, south-west of Camden. Open Monday to Friday 10 a.m. to 3 p.m., weekends, 9 a.m. to 5 p.m. Admission adults $4, children, concessions $1. The museum is located on a disused railway track outside of Picton and has a large collection of locomotives, old passenger cars, prison van, and railway memorabilia. A **steam train** travels from the museum to Picton and return several times a day. Admission for train and museum, adults $9, concessions $4.50. They also have regular steam train journeys out to the museum from Central Station, twice a month. The journey takes about 3 hours and costs $25 for adults and $15 for children, and a family of four costs $65.

Bus and Truck Museum (558 1234) Gannon Street, Tempe, off the Princes Highway. Open Sunday 11 a.m. to 4 p.m. Admission free. This museum has a large collection of buses of all ages shapes and sizes, lovingly restored by volunteers.

Sydney Tram Museum (521 7624) Lady Rawson Avenue, Loftus, 2 kilometres south of Sutherland, about a 1 kilometre walk south from Loftus station. Open Sundays, 10 a.m. to 5 p.m. Admission $5 for adults, children $3. A large collection of trams, many which used to run in Sydney before the last of the tracks were ripped out in 1961. Trams run down the short track at the museum.

Aircraft Museum (529 4169) 11 Stewart Street, Narellan (near Camden). Open Sundays. With 15 RAAF and Navy aircraft, this is a must for airplane buffs. You don't have to keep at arm's length from the aircraft and can watch the volunteers restore the planes.

Banking Museum (Westpac Museum) (251 1419) 6-8 Playfair Street, The Rocks. Open Monday 1.30 p.m. to 4 p.m., Tuesday to Friday, 10.30 a.m. to 4 p.m., Saturday 1 p.m. to 4 p.m. and Sunday, 12 noon to 4 p.m. Admission free. The exhibits trace the long history of this bank, and has displays of old coins and notes.

Historic Homes

Elizabeth Bay House (358 2344) Onslow Avenue, Elizabeth Bay. Open Tuesday to Sunday, 10 a.m. to 4.30 p.m. Cost $4 per adult, concessions $2. This lovely mansion dates from 1835 and was designed by John Verge.

Vaucluse House (337 1957) Olola Avenue, Vaucluse. Open Tuesday to Sunday, 10 a.m. to 4.30 p.m. Admission $4 for adults and $2 for concessions. This impressive estate was built in 1803 and is set in beautiful grounds that extend to the beach. They serve lunch and teas of interesting Aussie fare.

Old Government House (635 8149) Parramatta Park, Parramatta. Open Tuesday, Wednesday, Thursday, and Sunday, 10 a.m. to 4 p.m. Admission $4, concessions $2. It was built as the country residence of Colonial Governors beginning in the 1790s. It has a wonderful collection of early Australiana furniture and is set in lovely gardens.

Elizabeth Farm (635 9488) 70 Alice Street, Parramatta. Open Tuesday to Sunday, 10 a.m. to 4.30 p.m. Admission $3 for adults and $1.50 for concessions. Dating from 1793, it was the farm of the Macarthurs who pioneered merino wool production. The tea room serves lunch and teas.

Experiment Farm Cottage (635 5655) 9 Ruse Street, Parramatta. Open Tuesday, Wednesday, Thursday, Sunday, 10 a.m. to 4 p.m. Adults $3, concessions $1.50. The cottage of Colonial Surgeon John Harris in the 1820s, it was built on land from the first land grant in the colony, given to James Ruse who successfully grew wheat here. The cellar of the house has early farm implements and the house is furnished as a typical colonial gentleman's cottage from the mid-1800s.

Don Bank Museum (955 6279) 6 Napier Street, North Sydney. Open Wednesday 12 noon to 4 p.m. and Sunday 1 p.m. to 4 p.m. This four-roomed timber slab cottage was built in 1825. It is now a museum displaying the history of North Sydney.

Vienna Cottage (816 2255) 38 Alexandra Street, Hunters Hill. Open Saturday 2 p.m. to 4 p.m. and Sunday, 11 a.m. to 4 p.m. and by appointment. A National Trust property, this lovely cottage was built in 1871 and is typical of the old workers

Art Gallery of New South Wales

cottages in Hunters Hill. It features a display on the history of the Hunters Hill area.

Colonial House Museum (27 6008) 53 Fort Street. Open daily, except Saturday 10 a.m. to 5 p.m. Features period furniture in a Victorian House in the Rocks area.

Art Galleries

Art Gallery of New South Wales (225 1700) Domain, Art Gallery Road. Open Monday to Saturday, 10 a.m. to 5 p.m., Sunday, 12 noon to 5 p.m. Admission free, except for special exhibitions. Free bus Route 666 operates between Wynyard Station and the Art Gallery leaving Wynyard at 7 and 37 minutes past the hour and leaving the art gallery on the hour and half hour.

This impressive sandstone building was constructed in 1885. In 1988 the exhibition space was nearly doubled with an addition which seems to have lifted all the stuffiness out of the gallery. The art gallery has a vast collection spanning the history of art. Of particular interest is the Aboriginal collection and the Australian painters, especially around the turn of the century and into the 1920s. Try to catch one of the free guided tours. Licensed restaurant on Level 5 and a coffee shop which serves sandwiches on Level 3. There is also a good art bookshop with a large collection of prints, postcards, etc.

Museum for Contemporary Art (252 4033) Circular Quay. Located in the old Maritimes Services Board building, the art gallery is scheduled to open in 1991. Its collection is the most extensive modern art collection in Australia.

S.H. Ervin Gallery (27 9222) National Trust Centre, Observatory Hill. Open Tuesday to Friday, 11 a.m. to 5 p.m., Saturday and Sunday, 2 p.m. to 5 p.m. Admission adults $2, concessions $1. Exhibits on Australia's environmental and artistic heritage. Cafe and bookshop.

Manly Art Gallery and Museum (949 1776) West Esplanade (the harbour side), Manly. Open Tuesday to Friday, 10 a.m. to 4 p.m., Saturday and Sunday, 12 noon to 5 p.m. Admission by donation. Specialises in Australian art from 1900 to 1950 and it also has a small museum featuring beach memorabilia such as surf boards and bathing suits.

Commercial Art Galleries

There are a few dozen private galleries dotted around Sydney and the suburbs. The list provided here is not comprehensive but it does include the major galleries as well as a selection covering a range of different types of art. Most galleries are open Tuesday through Saturday. Visitors are most welcome just to browse. The galleries usually have two free brochures available which will help those wanting to tour the galleries—called *Museums and Galleries of Paddington* and *Museums and Galleries of Darlinghurst*. Look in the Saturday edition of the *Sydney Morning Herald* to see what's on at the galleries. The magazine *Art in Australia* provides the most comprehensive listing of gallery exhibits.

Macquarie Galleries (264 9787) 204 Clarence Street, Sydney is one of the outstanding galleries, with a long history dating from 1925. The gallery has worked closely with many famous Australian artists as their careers have advanced. Artists such as such as Rupert Bunny, Grace Cossington Smith, and William Dobell exhibited here. Exhibits leading contemporary Australian artists.

Australian Craftworks (27 7156), 127 George Street, The Rocks display their unusual craft items in the cells of this old police station.

The Craft Centre (27 9126) 100 George Street, The Rocks. Operated by the Craft Council of Australia, the centre has a big collection of leather, glass, ceramics, wood, jewellery, metalwork, and textile works, most of which are for sale.

Aboriginal Artists Galleries (250 7111) at the Sydney Opera House and also at 477 Kent Street (261 2929). These galleries have been established by the Australian government to market Aboriginal and Islander arts and crafts.

David Jones' Art Gallery (266 5640) 7th floor Elizabeth Street Store City exhibits a wide range of artists.

The Potters Gallery (331 3151), 48 Burton Street, Darlinghurst exhibits beautiful pottery and ceramics.

Robin Gibson Gallery (331 6692) 278 Liverpool Street, Darlinghurst is a well-known gallery housed in a lovely 19th century home. It shows paintings, sculpture drawings, and ceramics of leading Australian artists.

Garry Anderson Gallery (331 1524) 102 Burton Street, Darlinghurst promotes established and up-and-coming artists, mostly Australian.

Yuill/Crowley (698 3877) 270 Devonshire Street, Surry Hills specialises in particularly innovative modern art.

Watters Gallery (331 2556) 199 Riley Street, East Sydney displays a large collection of contemporary Australian artists.

Holdsworth Galleries (32 1364) 86 Holdsworth Street, Woollahra is a large established gallery, featuring well-known Australian artists.

Trevor Bussell Fine Art Gallery (324 605) 180 Jersey Road, Woollahra for Australian painters from 1800 to 1950.

Christopher Day Gallery (326 1952) Corner of Paddington and Elizabeth Streets, Paddington specialises in Australian 19th and 20th century Australian and European oil and watercolour paintings.

Josef Lebovic Gallery (332 1840) 34 Paddington Street is the place to go for prints and posters—a wonderful selection. They also have photographs.

Roslyn Oxley 9 Gallery (331 1919) 13-21 Macdonald Street Paddington often shows exciting and innovative works of new artists as well as the big names in Australian art. It is also active in presenting major international art in Australia.

Barry Stern Galleries (331 4676) 19-21 Glenmore Road, Paddington deals in both traditional and contemporary Austrlian art. He also has a second gallery with changing exhibits at 12 Mary Place, Paddington.

Hogarth Gallery (357 6839) 7 Walker Lane, Paddington. Has traditional Aboriginal bark paintings, carvings, antique artefacts, etc., as well as a range of contemporary art.

Artarmon Galleries (427 0322) 479 Pacific Highway. Artarmon has a large collection of early and contemporary Australian paintings and drawings.

Entertainment

Sydney has a tremendous variety of entertainment to suit all ages, cultures, moods, and budgets. This section gives a quick rundown on nightlife including pubs, clubs, performing arts, and film. For sporting events, sporting activities, and gambling, see other sections.

What's On

The daily newspapers are a good source of information for finding out what's on. The Friday *Sydney Morning Herald* has a good entertainment section (the blue pages) and their Saturday paper has several entertainment pages.

The free booklet called *This Week in Sydney* is available at most hotels and tourist places and gives a listing of entertainment on that week.

Another publication, containing information about the music scene in particular, is *On the Street*. Published weekly, it is available free from record shops.

Getting Tickets

Ticket agencies called **Ticketek** can be found all over the city, selling tickets for theatre, concerts, symphony, sporting events, and much more. Phone bookings can be made by ringing 266 4800. There is a recorded event information hotline: 11668.

You can get **half-price tickets** on the day of the performance by going to **Halftix** in Martin Place near Elizabeth Street. They are open Monday to Friday from 12 noon to 5.30 p.m. and Saturday 12 noon to 5 p.m. No phone reservations available. Halftix is also a Ticketek agency, so if you can't find what you're looking for at half price, the full range of normal tickets is also available.

Pubs and Bars

If you're looking for a cosy English style pub with olde worlde charm, the best place to go is down in the Rocks. The **Lord Nelson** at 19 Kent Street has its own brews of boutique beers. The **Hero of Waterloo** at 81 Windmill Street is another mid-19th-century bar full of character and customers. The **Harbour View** at 18 Lower Fort Street and the **Palisade** at 35 Beddington, just beyond the end of Argyle Street, are typical of Australian bars built early in the 20th century.

Other special spots are the **Marble Bar** at the Hilton on George Street, the **Gresham** at York Street, and the **Metropolitan** on the corner of George and Bridge Streets. The upmarket bars at the Regent and the Intercontinental Hotels are also popular spots. Wherever you are in the inner areas of Sydney, you won't have far to go to find a drink. In North Sydney there are several pubs and clubs catering to the office crowd, such as **Callaway's Real Ale Cafe** on Napier Street and **Sheila's** at 77 Berry Street. At Kings Cross of course there are plenty of pubs and clubs too. Oxford Street also has a variety of pubs and clubs and a few of these are gay pubs (see Gay Sydney below).

Music

This section gives a few leads on where to go for music, but check the newspapers and the weekly *On the Street*.

Opera/classical

For **Opera**, it's the Opera House of course, see Part 3. For classical music, the main venues are the Opera House Concert Hall, Sydney Town Hall on George Street or the Seymour Centre. The **NSW Conservatorium of Music** by the Botanic Gardens off Macquarie Street holds evening as well as pleasant twilight concerts on Fridays at 6 p.m. From time to time they also have free lunch concerts. Phone 230 1263.

Jazz

If it's jazz you're after there are several choices. **Soup Plus**, 383 George Street in the city (29 7728) is one of the most popular jazz venues and serves dinner as well in their cosy quarters. Bookings advisable. The **Berlin Hotel** in Cleveland Street is another popular jazz spot. There's also the **Real Ale Cafe** at 66 King Street with jazz from Wednesday to Saturday and the **Frying Times Fish Cafe** at 35c Ross St, Forest Lodge, Glebe (660 3302).

Clubs, Discos, Dance Halls

A recent craze in Sydney is the revival of the "Saturday night dance". Hordern Pavilion at the Showgrounds usually holds a dance on Saturday night and attracts huge crowds despite the $20-plus admission charge. There is generally no live music; people dance to the choices of top DJs instead. Sydney has dozens of clubs and discos; most of them have a cover charge, ranging from $5 to $15, and many serve meals. Most of them are open until 2 or 3 a.m. Check the entertainment section in Friday's *Sydney Morning Herald* to find out what's on in and around the clubs.

Klub Kakadu at 163 Oxford Street, Darlinghurst, is popular, with good shows. **Harbourside Brasserie** has a great location right on the harbour and draws large crowds: Pier One, Hickson Road, Millers Point. **All Nations** at 50 Bayswater Road and the **Air Crew Club** in the New Texas Tavern Hotel at 44 Macleay Street are good bets in Kings Cross. A few of the international hotels also have popular but pricey clubs. In the heart of the city the **Real Ale Cafe**, at 66 King Street, has an enormous beer list and good food as well as entertainment.

Theatre

Theatre opportunities are rich and varied in Sydney. The largest playhouses are The Opera House Theatre, The Theatre Royal on King Street, Her Majesty's Theatre on Quay Street, Railway

Square, and the Seymour Centre in City Road, Chippendale. But there are plenty of very good companies operating out of small theatres, often in lovely locations. Seeking out a suburban theatre and taking in a meal along the way makes for a fine night out. Here are a few suggestions but book ahead because these places are popular:

Northside Theatre Company (498 3166) 2 Marian Street, Killara.

The Wharf Theatre (250 1700) Pier 4 Hickson's Road, Millers Point has a great setting right on the harbour, with good restaurant and bar.

Ensemble Theatre (929 0644) 78 Mcdougall Street, Kirribilli (Milsons Point railway station) is another wharf side theatre situated on a quiet north shore bay.

Nimrod Theatre (699 6031) 500 Elizabeth Street, Surry Hills. There are several very good fringe theatre performances too such as at the **New Theatre**, 542 King Street, Newtown; the **Performance Space**, 199 Cleveland Street, Redfern; the **Genesian Theatre** at 420 Kent Street, City.

Cabaret/Comedy

Taking in a dinner and show at one of the cabarets can be a great evening's entertainment. Check the papers to see whether it is dancing girls and feathers, theatre cabaret, or comedy acts. Here are a few suggestions:

Parramatta Riverside Theatre, Church Street, Parramatta

Bankstown Theatre Restaurant (707 9766) Rickard and Chapel Roads, Bankstown.

Tilbury Hotel (357 1914) 22 Forbes Street, Woolloomooloo.

Kinselas (331 6200) Taylor Square, Darlinghurst.

Comedy Store (252 1480) off Jamison Street, City.

The Argyle Tavern Restaurant (27 7782), 12-18 Argyle Street, The Rocks. At the Jolly Swagman Dinner Show you will be served seafood and beef, damper, and pavlova and will be entertained by a cabaret including didgeridoo players, sheep shearers and folk singers.

Film

Most large suburban shopping complexes usually have a cinema or two. But the **two main cinema areas** in Sydney are on George Street south of Liverpool Street and on Pitt Street south of Market Street. These show the popular films, many of them American. If you can wait until a Tuesday night you can get in at half price. For international films, second-run films, etc., there are several very good cinemas around the city:

Dendy Cinema, MLC buiding, Martin Place

Valhalla Glebe, 166 Glebe Point Road, Glebe

Walker Street Cinema, 121 Walker Street, North Sydney

Academy Twin, 3A Oxford Street, Paddington

Gay Sydney

The centre of male gay Sydney is Oxford Street. The **Albury Hotel** at 6 Oxford Street, Paddington, is a favourite as is the

Flinders Hotel at 63 Flinders Street, Surry Hills.

Midnight Shift, at 88 Oxford Street, Darlinghurst, often has queues well down the street during its peak hour around midnight.

The **Jungle**, at 504 Elizabeth Street, Surry Hills, is a favourite women's club.

There is a 24-hour recorded message for "What's On" in the gay community. Phone 699 6320.

Poker Machines

One of the most astounding things about poker machines in Sydney is their ubiquity. No matter the suburb, no matter the type of club, you're bound to find poker machines. In Sydney there's no Las Vegas-style strip lined with poker palaces, floor-shows, and blazing neon signs. Australian poker machines are tucked away in clubs and if you don't know where to go, they can be pretty hard to find. Many of the larger clubs are affiliated with rugby league teams and the profits go towards supporting the teams. Club members, out of state and overseas visitors are permitted in these clubs. The clubs often have live entertainment, reasonably priced food and drink, as well as athletic facilites such as squash courts, gyms, and swimming pools.

The clubs are well worth a look inside if you've never visited one. They're truly a unique part of Australian social life. Here are some of the bigger ones:

South Sydney Junior Leagues Club, 558a Anzac Parade, Kingsford (349 7555)

North Sydney Leagues Club, 20 Abbott Street, Cammeray (92 6101)

Eastern Suburbs Leagues Club, 97 Spring Street, Bondi Junction (389 1011)

Blacktown Workers Club, Campbell Street, Blacktown (622 2355)

Panthers (Penrith) Leagues Club, Mulgoa Road, Penrith (047-31 5555)

Amusement Parks

Australia's Wonderland *By car, follow the Great Western Highway past Parramatta and turn south on Wallgrove Road. It's about 35 kilometres and a 45 minute drive from the city centre. By public transport there are buses from Fairfield, Blacktown, and Rooty Hill railway stations every half hour. Phone 675 0100 for bus times.*

If you're looking for a Disneyland in Sydney, this is the closest you'll get. It has stage shows, rides, games and a large shady picnic area. There is even "The Beach" which is not quite an accurate description of this enormous water complex of swimming pools with the highest and fastest water slides, one is six storeys high!

A great attraction is the Bush Beast, described as the largest, most breathtaking wooden roller coaster in the Southern Hemisphere. There's plenty of choice in fast food and restaurants. They even offer an Aussie-style barbecue of lamb chops, steaks, and sausages under the trees in Billabong Grove.

Wonderland is open on weekends, Saturday 10 a.m. to 9 p.m.

and Sundays 10 a.m. to 5 p.m. but closed from July to September. It is open every day during school summer holidays (Christmas to the end of January) and Easter break. There's a 24 hour telephone information service on 832 1777. Bring your swim suit. The admission price includes unlimited rides and all shows. Adults $18 and children $13.

Australiana Park is another popular spot for a family outing. About 60 kilometres west of Sydney, on Camden Valley Way, Narellan, it's a theme park with a train ride around the park, sheep shearing, dancing stallions, koalas and so on. Open daily from 10 a.m. to 5 p.m. and Friday and Saturday nights 7.30 p.m. to 12 midnight. Ring 606 6266.

Panthers World of Entertainment—its name says it all! This gigantic leagues club has expanded, and besides its poker halls, dining, and sports facilities it now has cable waterskiing, waterslides, and a car track. It's on the Mulgoa Road, (047) 31 5555.

The Harbour by Night

If you're taken by the beauty of the harbour at night, you might want to indulge in an evening harbour cruise. **Captain Cook Cruises** offer dinner and dancing under the stars on Friday and Saturday nights. The cost is about $50 per person. Reservations required, phone 251 5007.

To see the harbour at night at budget prices, State Transit offers a **Harbour Lights Cruise**, see Part 5.

Restaurants

Sydney offers a tremendous choice in restaurants and if you're willing to make short forays into the suburbs some of the ethnic eating is superb. Sydney's multicultural mosaic (more than one out of four Sydneysiders were born overseas) makes for an incredible variety of restaurants. The eastern cultures are very well represented with an enormous variety of Thai, Chinese, Vietnamese, Malaysian and Japanese and there's plenty of Indonesian and even Fijian. The Mediterranean cultures are very active on the food front with Italian and Greek restaurants dotted all over the city.

New South Wales' liquor laws permit many restaurants to be B.Y.O., meaning bring your own wine, beer, or other alcoholic beverages. Sometimes you see diners bringing in a crate of bottles, to get them through from sherry to wine, then port or cognac and beyond! Some restaurants charge a small corkage fee. When making a reservation check and see if they are B.Y.O. If you find yourself without drinks in a B.Y.O.—don't worry, there's sure to be a bottle shop nearby, and the pubs usually sell beer and wine to take away.

Food with a View

The extensive harbour and ocean coast offers dozens of possibilities for scenic, though usually expensive, eating. Here are a few:

Harbour Restaurant, Opera House Promenade Terrace (northern end), Opera House, phone 250 7581. You couldn't get much closer to the water without being on a boat at this seafood restaurant. There are indoor tables too. Specialties include Sydney rock oysters, balmain bugs, and crayfish. There's take-away fish and chips. About $30 per head.

Doyle's At the Quay The proprieters of Doyle's at Watsons Bay have opened this restaurant at Circular Quay, in the Overseas Passenger Terminal, phone 27 3400.

Doyle's, Watsons Bay They have two restaurants at Watsons Bay, called **On the Beach** and the **Wharf Restaurant**, phone 337 2007. A wonderful location near South Head with superb views up the harbour. It's easily reached by car or bus, or take Doyle's Water Taxi leaving Monday and Friday for lunch from Commissioner's Steps.

In the Rocks area there are several restaurants with views overlooking the Quay and the harbour. There's the **Italian Village** at 7 Circular Quay West (27 6111), a very large restaurant on three levels and an outdoor eating area on The Piazza. The **Waterfront Restaurant** at 27 Circular Quay West (27 3666) is another very large restaurant, with many tables outdoors under the masts and sails of a sailing ship. **Phantom of the Opera** restaurant at 17/21 Circular Quay West (27 2755) is heaven for wine buffs where the Australian Wine Centre has teamed up with this restaurant to offer fine food, wine tasting, and wine purchasing.

The **Sydney Tower** has two restaurants: Level 1 has *a la carte* menu and is pricier than Level 2 which has a fixed priced menu for a three course dinner. The Level 2 restaurant costs about $25 to $30 depending on the day and children are $10. There is no elevator charge to the restaurants. Ring 233 3722.

There are plenty of other possiblilites too. **Sails**, McMahons Point gives views looking back toward the city and the bridge. The **Wharf Theatre Restaurant** has a wonderful perch at the water's edge at Pier 4. A bit further afield there's **Manly Pier Seafood** restaurant with great views of the harbour. Right on one of the Northern beaches is the **Freshwater Restaurant**, Moore Road, Harbord.

Restaurant Strips

Browsing down the streets listed in this section will give a good choice of restaurants and cafes. The best bets at the budget end of the market are Chinatown, Glebe Point Road, Darlinghurst and King Street in Newtown.

Chinatown

Dixon Street has dozens of reasonably priced restaurants. The Food Centres such as downstairs in the Chinatown Centre on Dixon Street are particularly good value with meals around $5 to $6 (see Part 3). The **Marigold** at 299 Sussex is a favourite for *yum cha.*

Stanley Street and Crown Street, East Sydney

Just south of William Street you'll find a huddle of pleasant Italian restaurants catering for the budget and the middle market. There's **Bill and Toni's** at 72-74 Stanley Street, with good food upstairs and excellent cappuccino down. Immediately across from Bill and Toni's you'll see a shopfront with a pool table. If you walk through the pool room, past the cafe bar, and up the stairs you will have made it to one of Sydney's best budget restaurants. It's called **No Names** and specialises in serving good spaghetti at $4.50 a plate. They also serve a couple of other meat courses. Help yourself to the cordial and bread and cutlery if the waiters aren't so inclined. This is the place for a good quick feed, don't expect much service and don't linger too long. You can go downstairs for a cappuccino and gelato. Around the corner on Crown Street are the more upmarket but good Italian eateries such as the **Atlanta** at 41 Crown Street and **La Giardinettto** at 131 Crown Street, the latter having a pleasant outdoor eating area.

Oxford Street

Although this is a particularly good cafe strip, which also has dozens of restaurants, it can be rather seedy at night.

Glebe Point Road

About a 15-minute bus ride down Broadway from George Street, this strip caters to the university crowd and has a wide range of restaurants for price and type. It's a good street to restaurant-browse as most of the restaurants have a menu with prices posted outside. **Kim-Van's** at 147-149 serves excellent Vietnamese food. **B.J.'s** at 99 Glebe Point Point has an interesting international menu and is a favourite with vegetarians too. The Japanese **Rengaya** at 22 Glebe Point Road is good value with some unusual curry dishes.

Parramatta Road and Norton Street, Leichhardt

About 30 minutes by bus from the city along Parramatta Road, this is where you go for pasta and gelato and there are several good restaurants and cafes. **La Botte Dore** is a favourite, and they serve delicious seafoods on their special seafood nights. For sweets, there's **Mezzapica Pasticerria** at 130 Norton Street.

King Street, Newtown

About 15 minutes on the train from the city, this has a lot of good budget restaurants as well as some upmarket ones. There is an enormous variety of ethnic eating from Fijian to Turkish. **El Bhasash** is one of the most popular cafes and their desserts are delicious. There's **Pasha's Kebah House** at 490 King Street for good middle eastern food at good prices and the **Bali** for Indonesian at 135-137 King Street. **Mamma Maria's** Upstairs at 329a is a popular Italian restaurant and the **Fijiana** at 543 King Street has an unusual choice of Fijian/Indian fare. For Greek, there's **Glendi's** at 397 King Street with good food and a Greek band on weekends.

Darling Street, Balmain

By bus from city this is about 25 minutes, or for a pleasant stroll before dinner take the ferry to Darling Street wharf and walk up to the restaurant area. For curry, the **Lanka Curry House** at 235 Darling Street and for Greek, **Sofie's** at 374 Darling Street. **Jiyu No Omise** at 342 Darling Street is a popular Japanese restuarant. The bustling **London Hotel** at 234 Darling Street has good brasserie-style fare and Australia's best beers.

Military Road, Neutral Bay to Mosman

This is about 40 minutes by bus from Wynyard Station. In the last three or four years restaurants have sprung up everywhere along Military Road. Most of them cater to the middle price range, but there are budget and pricey ones here too. For Americans craving ribs or Mexican food, there's **Mischa's** for delicious ribs at 139 Military Road, Neutral Bay, and **Muchachos** with roll-your-own burritos at 130 Military Road. For eastern cuisine there's the **Bombay Bicycle Club** in Neutral Bay and the **Malay Curry House** at 161 Middle Head Road in Mosman. There are also plenty of cafes.

Manly

By ferry, Manly is about 25 minutes from Circular Quay. There is plenty of variety here, and Manly has many restaurants with good views of the harbour or ocean. The **Fairy Bower Tea House** has a range of pasta, fish, and meat dishes and tasty desserts and has great views too at 7 Marine Parade. **Manly Pier Seafood Restaurant** is at the water's edge on Commonwealth Parade. For a pub meal there is the **New Brighton** or the **Steyne Hotel** on the Corso. **Le Kiosk** specialises in seafood and has a wonderful setting at the edge of Shelly Beach.

Double Bay

Take a train to Edgecliff Station. This is the place for swanky, kosher, and late-night eating. **George's** and **"21"** are local favourites for the Double Bay crowd.

Ethnic Food

Here are a few hints if you're looking for a huddle of restaurants of a particular ethnic persuasion.

Spanish

Liverpool Street west of George. The **Sir John Young** pub on the corner has a cosy little Spanish restaurant with excellent paella. The **Spanish Club** at 88 Liverpool Street lets non-members in upstairs. The **Cafe Espana** across the street has food as well as excellent coffee.

Chinese

See Chinatown above, and in Part 3.

Vietnamese

Cabramatta, in the south-west of the city on the railway line, is where the Vietnamese live, and there is a host of restaurants here, especially on John Street. However, there are Vietnamese restaurants in all the restaurant strips described above if you don't feel like making the trek out west.

Thai

There is no particular concentration of Thai restaurants, but there are dozens scattered around the city. On Oxford, King, and Military Road you're sure to find a few.

Greek

The Greek restaurants are concentrated around Elizabeth and Liverpool Streets. There's the **Cyprus Hellene**, 150-2 Elizabeth Street, which serves a good value mixed platter with souvlaki, moussaka, and calamari, among other things. The **Hellenic Club**, 251 Elizabeth Street, is another pleasant Greek restaurant and the window tables offer nice views of Hyde Park. The **Diethnes** at 336 Pitt Street serves good lamb dishes in more than ample quantities.

Italian

You don't have to go far to find an Italian restaurant. Leichhardt and Newtown are good bets as well as Stanley Street in Darlinghurst (see restaurant strips above). **Rossini's** at Circular Quay is good value in a great location.

Fast Food

If you're after a chain of the McDonald's, Hungry Jacks, or taco variety—then George Street, south of Liverpool Street, is the place. For fast food with a bit more choice, Darling Harbour has several eateries. The numerous "milk bars" are good bets for fast food, most of them serve hamburgers, meat pies, fish and chips, chicken, milk shakes, and so on at good prices.

There are several good fish and chips places around the city. **Kirribilli Fish and Chips** is on Fitzroy Street, Kirribilli, and you can wander down to the harbourside park to enjoy your meal. Manly has a good fish and chips on the Corso, a few paces away from The Esplanade. At Bondi Beach you'll find good fish and chips and you can sit on the beach to eat them.

Meat pies are everywhere, but finding a tasty one is not easy. Try the food hall at David Jones' and the hot-bread bakeries rather than the typical snackbar variety. Harry's Cafe de Wheels at 1 Cowper Wharf Road, Woolloomooloo, is *the* late-night place for a pie or a hot dog.

Pub Eating

Often a good bargain can be had by stopping into a pub for a counter lunch. You won't get *haute cuisine*, but you'll get a square meal for $6 to $10. For the atmosphere of a pub and good food to match, try **Callaway's Real Ale Cafe**, hidden away on Napier Street, west of the Pacific Highway in North Sydney. They have delicious bagel sandwiches as well as a good range on the blackboard everyday. They serve breakfast too from Monday to Friday. The **Playfair Bar and Brasserie** on Playfair Street in the Rocks is another good eating, good drinking establishment as is the **London Hotel** in Balmain.

Club Eating

In Sydney there seems to be a club for every ethnic group and sporting group, as well as exservicemen's clubs. Most of these places serve reasonably priced food because they are subsidised by the poker machine takings. Intrastate and overseas visitors can usually get in without being accompanied by a member. The **Kirribilli Ex-serviceman's Club**, a few minutes walk from Milsons Point Station has good steaks and seafood plus a lovely view.

Where to Go for Breakfast

Breakfast or brunch in Sydney isn't so much of an institution as it is in North America. However brunch seems to be gaining popularity. **Callaway's** at 1-7 Napier Street, North Sydney, is open for breakfast from Monday to Friday. There's always **yum cha** (dim sum) in Chinatown, see Part 3. Many city cafes are a pleasant alternative for breakfast. **Le Croissant** at Town Hall station serves delicious croissant and cakes. **Phantom of the Opera** at 17/21 Circular Quay West is open for brunch on Saturdays and Sundays. Many pubs serve a hearty bacon and egg breakfast.

Shopping

Where to Shop

For shopping hours, see Part 1. The central business area of Sydney has a wide variety of shops. David Jones is the largest department store at Market and Elizabeth Streets and has everything a well-stocked department store should have. Grace Bros at George and Market Streets is the other large downtown department store. The Rocks and Darling Harbour area caters more for the tourist trade and there are some unusual speciality shops. The Queen Victoria Building has over 200 shops, many of them fairly pricey but it has a pleasant shopping environment and lots of choice. Double Bay is one of the most exclusive shopping areas with a wide range of fashions, and Australiana items. Military Road in Mosman is the sort of north shore equivalent to Double Bay, with quite a few designer shops. There are a few large suburban malls dotted around the suburbs, the largest are at Chatswood, Parramatta, Macquarie Centre in Ryde, Warringah Mall in Brookvale, Bankstown, Roselands and Eastgardens at Pagewood.

Souvenir Hunting

The Rocks, Darling Harbour, and the Queen Victoria Building are good places for souvenir hunters. Here you'll find cuddly koalas, Akubra hats, T-shirts, opals, sheepskin products, and so on. For a wide selection in Australian bush fashion, like stockmen's coats ("drizabones") and Akubra hats, there's Morrison's at The Rocks. For a sachibukoro (kangaroo skin coin purse) and other quality souvenirs, go to Koala Wine Estates,

Shop 15, Level 2, Dymocks Building, 428 George Street. The Argyle Centre and the Craft Centre in The Rocks are good places for crafts. The gift shops at some of the attractions often have interesting Australiana items and books, such as the Barracks gift shop on Macquarie Street, the Opera House gift shop, the Australian Museum, the Taronga Zoo gift shop, and the Power-house museum. For outrageous surfboard shorts and other swimming gear, the shops around Manly and Bondi are good bets.

Duty-free Shopping

As well as at the airport you will see duty free shops dotted all around the city. You need an international airline ticket to buy in these shops and you cannot open your package until you pass through customs on your departure. You cannot take the goods out of the shops more than 10 days prior to your flight but many shops deliver to the airport. Good buys can be had for film, cameras, electricals, hard liquor, perfume and fashion items but shop around and bargaining does go on. The duty-free shops at the airport are considerably more expensive (especially for liquor and cigarettes) than in the city, so try to stock up before your departure.

Outdoor Gear

There are several good shops offering a wide range in sleeping bags, day packs, rucksacks, camping gear, ski equipment, and outdoors clothing. Most of the shops are clustered around Kent and Sussex Streets. Paddy Pallins at 507 Kent Street is one of the largest. On Sussex Street you'll find Wildsports at No. 327 Sussex, and Mountain Equipment at No. 291. If you find this quality gear too expensive you can buy cheaper packs and sleeping bags at one of the army surplus shops on Pitt Street or George Street south of Liverpool Street.

Sports Gear

The beach suburbs such as Manly, Bondi, and Mona Vale have plenty of shops for windsurfers, surfboards, wet suits, and the like. There are also a few on Military Road in Mosman. For surfboard and windsurfer rentals see the section below. The chain store Mick Simmons has a wide range of sports gear, including skin and scuba gear.

Books and Records

Dymock's at 424 George Street is one of Sydney's largest bookshops. For academic books plus a wide variety of every-thing else there is the Co-op Bookshop on 80 Bay Street, off Broadway. A fine central city bookshop is Abbey's at 131 York Street. For a shop with atmosphere try the New Edition Book-shop at 328 Oxford Street, Paddington. At 191 Glebe Point Road there is Glee Books open until 9 p.m. every night. On the North Shore there is the Constant Reader at 18 Willoughby Road, Crow's Nest. For art, architecture and the like there is Lamella Books at 333 South Dowling Street, Darlinghurst. There are a large number of discount bookshops which often offer good bargains, especially on picture coffee-table books. Books Un-limited and Colourcode are two of the large chain discount book shops. For travel guide books and maps, the Travel Bookshop, with their main shop at 1 Jamison Street near Wynyard Station off George Street, is the best bet.

For records, tapes, and compact discs there is the Brashs chain with a shop at 244 Pitt Street. There's Palings at 143 York Street and the Virgin Records shop at Darling Harbour. Folkways at 282 Oxford Street is an excellent music shop.

Markets

Sydney has several interesting markets. Besides the shopping attraction most of the markets offer food and refreshments, and some have musical entertainment.

Paddington Village Bazaar, Oxford Street, is held each Saturday and sells things like second-hand clothes, crafts, and books. It is a large, pleasant market with more than 250 stalls.

Balmain Market, St Andrews Congregational Church, Dar-ling Street. This market has a wide variety of food, craft, and second-hand goods.

Flemington Markets at Flemington railway station on the western suburbs rail line is the gigantic produce market but they also have a wide range of other goods. Open Friday 11 a.m. to 4 p.m., Saturday, 6 a.m. to 1 p.m. for fruits and vegetables only and Sundays 9.30 a.m. to 4.30 p.m. They can be reached by rail from Flemington Station.

Paddy's Markets have had a long history in Sydney. For decades they occupied the market place at the Hay Market and offered a wonderful range of produce, souvenirs, clothes, and other products. Real estate prices have driven them out of Central Sydney and now they're located at Flemington Markets on Friday and Sunday (see above) and at Redfern on Saturday and Sunday, right near the Redfern railway station, one stop past Central Station.

Fish Market, Gipps Street, Pyrmont. A trip to the fish market is a popular weekend activity for Sydneysiders. Open everyday from about 6 a.m. Besides getting fish at good prices, it's entertainment. There's an enormous variety of fish and seafood ranging from whole tunas to giant green prawns. Noise levels are high with lots of shouting, sloshing of ice, and there's even a few mini cement mixers twirling octopus around to get rid of the ink.

Sports and outdoor activities

Sydney is a sporting person's paradise, the weather is ideal and the choice limitless. And generally the costs to rent equipment and participate are very reasonable. The Yellow Pages are a great source for listings of all the clubs and facilities but make sure you look under "C" for Clubs if you cannot find the listing under the sport.

On the Water

Sailing and boat hire

There are plenty of sailing schools around the harbour offering lessons and you can also rent lasers, yachts, and outboards. Waltons Hire Boats, 2 The Esplanade, Balmoral Beach (969 6006) has everything from yachts, sailing dinghies, windsurfers, canoes, surf skis, and aluminium fishing boats. You can find calm water here but you'll have to share it with many others on the weekend. For beginner sailors it can be a bit tricky weaving your way out among the moored boats. Located on Middle Harbour, it is a lovely stretch of water to explore. The adventurous can go through the Heads into the main harbour, or make a trip to Manly.

If you'd like to crew in one of the many sailing races, you can join one of Sail Australia's harbour races on Saturday mornings. Cost is $25 and no experience is necessary (See address and phone number below). For experienced sailors, one possibility is the Cruising Yacht Club at Rushcutters Bay (326 2399).

Australian Sailing School, The Spit, Mosman (960 3077) learn to sail craft from yachts and dinghies to sailboards.

East Sail, New Beach Road, Rushcutters Bay, (327 1166) lessons, crew positions, and yacht hire.

Sail Australia, 23a King George, Lavender Bay, North Sydney

(957 2577). Located just west of the harbour bridge, they offer lessons and yacht hire.

Rose Bay Outboard Hire, Vickery Avenue, Rose Bay, (326 1507)

Cronulla Houseboats, 1-3 Tonkin Street, Cronulla, (523 6919). They rent houseboats, half-cabin and fishing boats, and are conveniently located behind the Cronulla railway station.

For houseboat rentals on the Hawkesbury River, see Part 7.

Fishing

The possibilities are limitless. You can hop on a ferry and get off where you see others fishing, at Cremorne Point or Ashton Park for example. You can rent a boat at several places along the waterways (see above for a couple of suggestions or go up to the Hawkesbury River). Alternatively, you can go on a fishing charter. Ring Manly Sports and Dive on 977 4355.

Surfing

The best bet for surfboard hire is Surfboards Manly Style at 6/49 North Steyne Street (977 3363). They'll rent you a boogieboard or a malibu, as well as the conventional surfboards.

Sailboarding/windsurfing

Renting a sailboard will cost about $15 an hour plus a bond. Many places will rent them long term or sell it to you on a guaranteed buy back basis. The most popular quiet water windsurfing spots are Rose Bay, Balmoral, Manly Cove, Narrabeen Lakes, and the Pittwater side of Palm Beach. For wavejumpers Long Reef and Palm Beach are popular. Here are some sailboard rental places:

Rose Bay Windsurfer School, 1 Vickery Avenue, Rose Bay (371 7571).

Balmoral Marine, 2 The Esplanade, Balmoral Beach (969 6006).

Long Reef Sailboard and Hire, 1012a Pittwater Road, Collaroy (982 4829) for wavejumpers and sailboards.

Scuba diving

There are some interesting dive spots in the harbour and off the ocean beaches but visibility is often poor, especially in the harbour. Generally winter diving in Sydney is better because the water is clearer. Whale Beach Caves on Whale Beach, one of the northern beaches is a popular, protected spot. Long Reef and Shiprock are other favourites. Many Sydney-siders make a day trip or weekend trip either north to Seal Rocks or Jervis Bay south of Sydney. The big attraction at Seal Rocks is the reef, about 3 kilometres from shore which has interesting sea life as well as a few shipwrecks. Jervis Bay has beautiful white beaches and normally clear waters. There are interesting rock formations and caves as well as reefs. Many of the dive shops organise package trips at reasonable rates to these spots.

Several dive shops also offer scuba diving courses lessons. Pro Dive at Manly (phone 977 4355), Pro Dive at Coogee (phone 665 6334), or Frog Dive Scuba Centres (phone 958 5699) are all good bets for equipment hire, lessons, and excursions.

Swimming

See the description of the harbour areas, Bondi and southern beaches, and the northern beaches to pick an ocean or harbour beach spot.

For those who prefer a swimming pool, there's lots of choice. Boy Charlton is a pleasant 50-metre pool in the Domain, just up from the Art Gallery. The North Sydney Olympic Pool has a harbourfront location and is one train stop north of Wynyard at Milsons Point. Many of the beaches also have seawater pools or swimming enclosures such as Bondi, Balmoral Beach, Freshwater, Neilson Park, and Redleaf Pool in Double Bay.

On the Land

Golf

Sydney has a very large number of courses scattered around the city. Many of them are open to the public all the time, some have restricted hours for non-members, and the private clubs often let visitors with an overseas membership play. The cost of playing at most public courses is very reasonable. For a driving range there is Moore Park, open in the evenings Monday to Friday, ring 663 3960. Here are a few suggestions:

Moore Park, Cleveland St, Moore Park (663 3960) This is a pleasant course and easy to get to by bus. They rent clubs and visitors can play anytime during the week and after 12.30 p.m. on weekdays. Half price rates for green fees and club rentals apply after 4 p.m.

St Michael's, Jennifer Street, Little Bay. Don't let the great ocean views distract you. Unrestricted playing times on Monday, Wednesday, and Friday, limited hours on the other days.

Northbridge at Sailor's Bay Road, Northbridge (958 7517) and Lane Cove on River Road, Lane Cove (428 1316) are short courses on the north shore and easy to get to.

Eastlakes, Gardeners Road, Kingsford (663 1374). Playing times anytime Monday to Friday, limited times on weekend.

Tennis

Courts are around $6 per hour during the day and $10 at night. Here are a few suggestions:

Palm Tennis Court Hire, Trumper Park, Quarry Street, Paddington (32 4955). Open day and evenings every day, six synthetic grass courts.

Moore Park Tennis Courts, Anzac Parade, Moore Park, (662 7005). Open day and evening every day, ten courts.

North Sydney Tennis Centre, 1a Little Alfred Street, North Sydney (957 1877).

Manly Tennis Centre, Raglan Street, Manly (977 3159), Open day and evening every day, six synthetic courts.

Tennis World, 16 Epping Road, North Ryde (888 7466). Open day and evening every day, eleven courts.

Horse riding

For a horse ride in the middle of the city, Centennial Park offers a 4.5-kilometre ride on a track around this lovely park. They leave each hour between 9 a.m. and 4 p.m. and cost around $12 per hour. It's a very popular activity so reserve ahead if you want to ride on a weekend.

Centennial Park Horse Hire, Driver Avenue, Centennial Park (332 2770).

Blue Ribbon Riding School, The Showgrounds (33 3859)

Challis Riding School, 69 Darley Road, Centennial Park, (398 4879)

There are plenty of other places to ride but you'll have to go to the outer suburbs or if you're really big on horses a trip to the Southern Highlands or the Blue Mountains (see Part 7) offer lots of choice, lovely scenery, and overnight trips. A couple of suggestions in the outer suburbs are Palomino Corral at Belrose (450 1726) and Cattai Park, a large picnic ground on the Hawkesbury River (045-77 5045).

Cycling

Several bicycle hire shops can be found around Centennial Park (see Part 4 for a description of the park). Most are open every day.

Centennial Park Cycles, 50 Clovelly Road, Randwick (398 5027).

Australian Cycle Company, 28 Clovelly Road, Randwick (399 3475).

The Red Bike Shop, 11 Clovely Road, Randwick, (399 9060).

In Parramatta Park (see Part 4) there's Parramatta Boat and Cycle Hire, Riverside Drive (683 5905).

Indoor Sports

For squash courts, check the *Yellow Pages*. There are also lots of gyms and aerobic centres, also listed in the *Yellow Pages* under Health and Fitness Centres. A couple of centrally located centres are Aerobic City, 327 George Street, entrance on Wynyard Lane, phone 29 6632, or Healthland, 40 Miller Street, North Sydney 922 2155.

In the Air

If you want to see Sydney and environs by air, there are lots of choices, including helicopters, air ships, hot air ballons, even tandem skydiving and, of course, planes.

Aquatic Airways (919 5966) have scheduled flights between Rose Bay, Palm Beach, Gosford, and Newcastle as well as scenic flights.

Hazelton Airlines (235 1411) offers the "The Top Ten" Skytour with a flight over the harbour, northern beaches, Pittwater, Hawkesbury River and the Blue Mountains. You do all this in 70 minutes and it costs about $110.

Sydney Aerial Tours (709 4953) offers daily flights around Sydney.

Water Wings (92 0272) will take you on their floatplane up to Pittwater or drop you off at a secluded restaurant on the Hawkesbury River.

Red Baron Scenic Flights (771 1333) offer flights in a sopwith camel, aviation hood and all.

Sydney Helicopter Service (27 5151) offer daily flight around Sydney, the Hunter and the Blue Mountains.

Skycruise (221 3688) offers airship flights over Sydney. Price around $150.

A few companies offer hot air ballooning in the countryside.

Camden Valley is the closest location and is about 60 kilometres from Sydney. Other popular places are the Hunter Valley, Canowindra and Canberra. The cost is about $100 per person per flight and dearer if transport to the ballooning site is required.

Airborne Adventures (063 44 1717) offers ballooning weekends in Canowindra.

Balloons Aloft (607 2255) Flights at Camden with pick-up in the city. Also flights in the Hunter Valley and Canberra.

Balloon Sunrise (818 1920) Flights in the Hunter Valley and Camden Valley with city pick-ups.

If you have an urge to try skydiving, the uninitiated can go tandem skydiving safely strapped to a qualified instructor. Only takes 20 minute instruction prior to your free fall from 2000 metres. It costs $220 for the jump, and action photos are extra. Ring Tandem Skydiving on 597 5918.

Spectator sports

As well as being a haven for the participating sporting types, Sydney presents a wide and interesting choice for the spectator.

Cricket and rugby league football are the two most popular spectator sports. An afternoon of cricket or football is a great sporting and cultural event for overseas visitors.

Cricket

The cricket season runs from December to March. The matches are at the Sydney Cricket Ground, Moore Park and it's an easy bus trip from the city. The stadium seats 40,000 people. The highlight of the season is the international test matches against the other cricketing countries such as England, West Indies, and India. A test match may take up to five days. For the uninitiated a "limited over" one-day game may be the best idea, either in the day or an atmospheric night game under the lights. The intrigue of this game becomes apparent after a proper introduction, but if you're North American, whatever you do, don't compare the game with baseball or you'll send your Australian mates running for cover.

Football

There are four different sorts of football (codes) played in Sydney: soccer, Rugby League, Rugby Union, and Aussie Rules. The season runs from late February to the grand finals in September. At every sports ground around the city you'll see some sort of football being played all Saturday long. Rugby League is the most popular and Sydney-area teams battle it out in their suburban grounds. The grand final is held at the Sydney Football Stadium at Moore Park. Aussie Rules is an exciting game—if you're going to Melbourne, catch a game there in the heart of Aussie Rules country, otherwise watch the schedules for a Sydney Swans home game at the Sydney Cricket Ground.

Basketball

Australia has a National Basketball League. It's not up to the American pro standard but it's good quality basketball and there are quite a few American imports playing in Australia. The Sydney Kings play at the Sydney Entertainment Centre in Chinatown and the season runs from May to October.

Sailing Races

By far the most exciting race on water is the 18-footers, the fastest of all sailboats. Their season runs from September to mid-March and they race on Saturday and Sunday afternoons. Sailed with a crew of three, they have giant trapezes that defy gravity. You can watch them from ferries that follow the race and even place a bet. Phone 32 2995. The best harbour shore viewing points are Ashton Park or Cremorne Point.

Surf Carnivals and Surfing

Surf carnivals are held at many of Sydney's beaches on weekends from October to March. They're usually very colourful spectacles with surf-boat races and surfboard riding events as well as rescue demonstrations and marching. Ring the Surf Lifesaving Association on 597 5588 to find out what's on.

Surfing competitions are held mostly in the summer months. The location often depends on conditions on a particular day.

Horse Races

Horse races are very popular in Sydney. All four racecourses (Randwick, Canterbury, Rosehill and Warwick Farm) are accessible by public transport. Saturday is the big day at the races but there are also races every Wednesday at one of the courses. Trotting meetings are held on Tuesday and Friday at Harold Park in Glebe. If you prefer to watch the greyhounds, then Wentworth Park in Glebe is the place to go. Check the newspapers for events, location, and time.

Tennis

The major tournaments are usually held at White City in Rushcutters Bay or at the Sydney Entertainment Centre in Chinatown between October and February. Check the papers or ring tourist information.

Further reading

If you want to read more about Sydney's built and natural environment, past and present, there are many excellent bookshops that could help you. However, it may be best first to pay a visit to a municipal library, or to the State Library, as quite a few good books on Sydney are now out of print.

If you don't have much time, a quick browse through *The Australian Encyclopaedia*, published by the Australian Geographic Society, may satisfy at least some of your curiosity.

If you feel that the only way to see a city is on foot, the Department of Planning (NSW) issues several walking tour booklets, illustrated with lovely pen and ink sketches, whilst walking tour pamphlets are available from Parramatta, Strathfield, North Sydney, and Hunters Hill Council, as well as from the Balmain Historical Society.

History

Birch, Allan and David Macmillan, *Sydney scene—1788-1960*, Hale & Iremonger, Sydney 1962

Kelly, Max, *Nineteenth century Sydney: essays in urban history*, Sydney University Press, Sydney 1978

Roe, Jill (editor), *Twentieth century Sydney: studies in urban and social history*, Hale & Iremonger, Sydney 1980

Spearritt, Peter, *Sydney since the twenties*, Hale & Iremonger, Sydney 1978

Willey, K, *When the sky fell down: the destruction of the tribes of the Sydney region*, Collins, Sydney 1985.

Architecture and Town Planning

Webber, Peter (editor), *The design of Sydney*, 1988.

Haskell, John, *Haskell's Sydney*, Hale & Iremonger, Sydney 1983

Coltheart, Leonie, *Significant sites: history of public works in New South Wales*, 1990

Spearritt, Peter and Christina DeMarco, *Planning Sydney's future*, Allen & Unwin, Sydney 1988

Guides

Gunter, John, *Sydney by ferry and foot*, 1987

Hughes, Roland, *Wraps-up Sydney*, Roland Hughes publishing, Sydney 1987

Proudfoot, Helen, *Exploring Sydney's west*, 1987

Spence, Peter, *Sydney by public transport*, Gary Allen, Sydney 1989

Thomas, Tyrone, *100 walks in New South Wales*, Hill of Content, Melbourne 1977

Food

Dunlop, Lisa, *A taste of Sydney—a guide to Sydney's speciality food shops*, 1988

Gregory's cheap eats in Sydney

Mayne, Robert and Gwen, *Pocket Australian Wine Companion*, 1988

Schofield, Leo, *Sydney Morning Herald good food guide*, Anne O'Donovan, Melbourne

Field Guides

Pizzey, Graham and R Doyle, *A field guide to the birds of Australia*, Collins, Sydney 1980

Allan Fairley, *Field guide to the national parks of NSW*, Dent, Melbourne 1978

Index